THE RELUCTANT SURVIVORS

THE RELUCTANT SURVIVORS

A FAMILY GUIDE TO THE PREVENTION AND TREATMENT OF RADIATION SICKNESS

Wayne D. LeBaron

Dream Garden Press
Salt Lake City
1984

To my wife, LaWana

Copyright © 1984 by Wayne D. LeBaron and Dream Garden Press.

All rights reserved. No portion of this book may be used or reproduced by any means, with the exception of limited quotations for review or scholarly purposes, without written permission from the publisher.

ISBN: 0-942688-15-5   First edition.

Library of Congress Catalog Card Number: 84-70567

Manufactured in the United States of America.

Dream Garden Press • 1199 Iola Avenue • Salt Lake City, Utah 84104

PREFACE

THOUGH intended to be as thoroughly researched and comprehensive as possible, this book does not pretend to have all the answers. In addition to sources mentioned throughout the text and collected as references at the ends of the chapters, a bibliography is included which can guide the interested person to materials which can help answer remaining questions. It is understood that not every alternative substance and therapy is mentioned, nor that it is necessarily desirable to do so, if it was indeed possible. A program, to be effective, must be capable of being carried out. Thus, the criteria employed in the compilation of this book have been *availability*, *ease of administration*, and *affordability*. It is to be hoped that members of the medical profession and community at large who can improve upon suggestions made here will raise their voices in common concern.

One of the most disturbing aspects of the societal malaise induced by ignorance and fear of the effects of thermonuclear war is the conviction on the part of so many that "they don't want to think about it." No matter where a person stands on the great issues of our time, he does not have the option of advocating ignorance. Describing the *reality* of life after such a cataclysm may increase the likelihood that no one will ever have to actually experience it. It is with that hope that this book has been written.

CONTENTS

PREFACE	V
INTRODUCTION	1
PART I	4
1. A BAD BEGINNING	5
2. CRISIS RELOCATION PLAN: FACADE OR FACT?	8
3. EMERGENCY MEDICAL CARE: ILLUSION OR REALITY?	10
PART II	14
4. NUCLEAR PHENOMENA	15
Electromagnetic Pulse (EMP)	15
Flash Blindness	18
Black Rain	19
5. FALLOUT AND NUCLEAR RADIATION	20
Types of Radiation	22
Radiation from Materials Deposited Internally	23
Protection Against Radioactive Fallout	24
Suggested Activities Between Attack and Arrival of Fallout	25
6. PROTECTION OF FOOD AND WATER	28
Food Protection and Decontamination	29
Water and Water Supplies	31
Water Source Safety	34
Purification Methods	35
7. DECONTAMINATION	37
Procedures for Inanimate Objects	39
Protecting Individuals	41
8. THE ACUTE RADIATION SYNDROME (RADIATION SICKNESS)	44
Radiation-Injured Tissue Regeneration	47
Cell Division and Tissue Damage	48
Enzyme Inactivation	49
Clinical Course of the Illness	50
9. PRE-EXPOSURE PREVENTION AND TREATMENT	57
Optimum Health	57

Immunity and Immunization 59
Substances Beneficial in Prevention of Damage
 by Ionizing Radiation 60
L-Cysteine 61
Potassium Iodide 63
Vitamin C (Ascorbic Acid) 67

10. TREATMENT OF RADIATION SICKNESS 71
Practical Considerations 72
Continuation of Preventive Medications 72
Infection and Resistance 73
Antibiotic Therapy 74
Tetracyclines—Terramycin and Aureomycin 75
Anti-Emetics 77
Vitamin Supplements 78
Amino Acids 79
Pain Relievers 81
Fluid and Electrolyte Balance 82
Nutrition 83
Patient Diets 83
Dietary Management for Patients Suffering Radiation Injury 86
Recommended Foods for Radiation-Injured Patients 87
Nursing Care 87
Rest 89

APPENDIX 1: Half-lives and Types of some Radioisotopes ... 91
APPENDIX 2. Forms to Record Possible Symptoms
 of Radiation Sickness 93
NOTES 96
SELECTED BIBLIOGRAPHY 100

The Reluctant Survivors

INTRODUCTION

THERMONUCLEAR war, resulting in mass annihilation, is often assumed to be the inevitable destination of our species. In many people this assumption has resulted in complete resignation, an attitude of helplessness. The malaise and apathy so characteristic of industrial society must in part be explained by this threat, which lurks behind each international incident, and just as frighteningly, can become real through commonplace realities like computer failure, operator error, or personal miscalculation. The result is that we often act as if the bombs have already dropped and we are simply awaiting vaporization. But we are *not* dead yet. And it is incumbent upon those still concerned with the future that they recognize this fact.

In late 1983 a great deal of interest in this subject was generated by the ABC Television special, *The Day After*, which graphically portrayed a scenario of thermonuclear war and its immediate aftermath. Therein, many of the issues dealt with in this book were explored. A common response by viewers reflected their hopelessness, expressed in the statement, "I hope I go quickly."

Yet it is probable that significant numbers of Americans would survive a nuclear strike, though it would unquestionably be an unprecedented disaster. What if you (and your family) are among *The Reluctant Survivors?* This book is an attempt to provide the lay person who chance or foresight spares a means to help himself in the post-war period.

Although most people are profoundly ignorant and fatalistic regarding radiation—the dreaded by-product of a nuclear reaction, radiation sickness (Acute Radiation Syndrome) is among the best scientifically understood aspects of thermonuclear war. The survivors of Hiroshima and Nagasaki have probably been the most exhaustively studied populations in medical history. In addition, significant research has been conducted on accidental radiation sickness victims. There exists a vast medical literature on prevention and treatment of the malady, as well as thoroughly documented case histories. It is unfortunate that so little of this information has been disseminated to a population that may eventually desperately need it. This results in part because radiation sickness is not encountered every day, even by health care professionals. It is regrettable though perhaps unavoidable that few are even superficially trained to deal with it. It is probable that in the long run a worst case scenario will be played out in a nuclear power plant, and people living nearby will be victimized by various levels and types of radiation. Though that event will likely resemble the incident at Three Mile Island in that it will be disaster in slow motion with the whole world standing impotently by, it will still overwhelm civilian and military planners with unanticipated mass hysteria. Only individuals competent to deal with the dangers *on the family level* will be able to enhance their chances for survival.

What makes worse the failure to get this information to the public is that it is applicable in other, less controversial areas as well. For example, many of the substances suggested in chapter six are commonly employed to protect cancer patients receiving radiation treatment from its damaging effects. Methods of blocking the thyroid gland have been practiced for decades, and are used in some places prior to taking simple X-rays.

But the primary thrust of this book is to provide a usable plan to the cautious lay person in case of thermonuclear war. In that event, conventional medical practice will cease and dependence upon family members will become absolute. Though this is not another "dig a hole in the ground and bring food, water, and blankets" manual, it does assume that thoughtful people have recognized the danger and stockpiled essentials for that eventuality. As part of a complete program, this book provides a clear, concise statement of what types of radiation are likely to be encountered, their biological effects, and what can be done to prevent and reduce damage. It relies upon two underlying assumptions: 1) Substances and procedures are available which can effectively protect people from the destructive effects of ionizing radiation. This refers primarily to drifting radiation as opposed to the massive levels very close to a nuclear detonation, and 2) Radiation sickness can be treated relatively successfully at the family level, if individuals learn the proper procedures and obtain the necessary materials, among other simple preparations. But there is another, perhaps more important, assumption also being made: that the goal of reduced damage from radiation is desirable; that we are indeed *still alive*. There is a caricature of the "survivalist" many of us have in our minds. He is paranoid, pessimistic, selfish. But the other side of his obsessive preparedness is knowledge that, *if* the time comes, anybody around to worry about it will wish that he, too, had been storing nuts. These concerns are not ideological, not bound up in a naive political doctrine of isolationism. They are a minimal response to a frightening potential reality, and perhaps a way of reducing the bewildering and maddening acquiescence of a population doomed as much by their own inactivity as by belligerent technology gone mad.

PART I

THE following pages deal with the dilemma we face, how it developed, and its implications. First, we look at the beginning of the nuclear age, the use of atomic weapons against Japan at the end of World War II, the subsequent development of thermonuclear (hydrogen) bombs, and the precarious balance of terror between the superpowers that has prevailed since the early 1950s. Then we discuss the history of "civil defense" in this country, and the implications of our lack of preparedness. We conclude with a look at the health and medical care system of our country in an effort to evaluate what will remain in event of thermonuclear catastrophe.

1. A BAD BEGINNING

ON August 6, 1945, the era of nuclear war began.
Hiroshima: 7:09 A.M.—The air raid siren at Regional Military Headquarters announced another day of war. An American bomber had been sighted on course toward the city. Most of the inhabitants were not particularly concerned and did not even seek cover. After all, American planes had been flying over the city from time to time, but had never attacked. It was always a relief to see them continue on to another target. Most often one or two would fly over, the long range bombers being capable of flying the distance from Tinian Island and returning. The Americans always sent a reconnaissance plane first, to check the weather, and to let waiting bombers know if it was suitable to drop their loads of destruction on the homeland.

The all-clear soon sounded. There was only one B-29, and it had turned away from the city. It could be seen as a speck in the sky at nearly 40,000 feet. How much damage could one B-29 cause to a city of over 340,000 people? Furthermore, it was turning away from Hiroshima, and soon was out of sight.

Inside the B-29, the radio operator was transmitting a message to its sister aircraft which would be over Hiroshima about an hour later. The message: "Fair weather over target, ready for raid." The city's people busied themselves with their daily tasks.

At 8:15 A.M. two weather observation B-29's accompanying a third, nicknamed the "Enola Gay," flew over. A few seconds later the entire city was enveloped in a blue-white glare of light, followed by intense heat, a heavy shock wave, and a terrific wind. One hundred thousand human beings were suddenly gone, along with most of the city. Three days later the scene was repeated at 11:02 A.M. at Nagasaki, a city of 272,312 people.

Publications on every aspect of this tragic, yet brief, nuclear war are available. The use of atomic weapons against Japan and subsequent testing has provided a great deal of information to the scientific community. During the last thirty-nine years the fortunes of war and politics have moved swiftly. Mankind has not yet learned to live with nuclear weapons, and, as Herman Kahn put it in 1960, has learned even less about "conducting international relations in a world in which force tends to be both increasingly more available and increasingly more dangerous to use."[1]

The implications of using nuclear weapons were not fully known in 1945, and that they were deemed "thinkable" is evidenced by their actual use. With continued testing, development of the hydrogen (thermonuclear) bomb, expanded research on both short-term and long-term effects, and the building and stockpiling of nuclear arsenals, their use became "unthinkable." Strikingly more sophisticated weapons have since been stockpiled, along with space-age delivery systems. This process has contributed to the notions of deterrence and mutually assured destruction (MAD) which have dominated strategic planners for thirty years. These ideas rest upon a notion of preparedness, such that if, again according to Kahn, "both opponents are prepared the old-fashioned distinctions between victory, stalemate, and defeat no longer have much meaning."[2]

In the very recent past, the idea of a "winnable" nuclear conflict has been revived. It was announced in August 1982 that the Pentagon had completed a master plan designed to allow the United States to win a protracted war against the Soviet Union. This is predicated on the use of "tactical" nuclear weapons, those of somewhat smaller mega-tonnage, which can be delivered in the context of a limited war, i.e., to perform surgical strikes on an opponent's primary industries, ports, etc. Many view the advent of such thinking as counter-productive, making it that much more certain that nuclear weapons will be introduced, and that once used, escalation will be unavoidable. Robert Scheer, in his book, *With Enough Shovels*, neatly summarizes this thinking: "Increasingly accurate missile technology and sophisticated means of communications have produced the confidence in some quarters that nuclear war need not be fought as one spasmodic episode with little but radioactive rubble to show for the effort."[3]

The strategic warheads that tip today's ICBMs are usually between thirteen and thirty-five times more powerful than those dropped on Japan. Many are several times more destructive than all the firepower of World War II. Contrary to the reassuring nonsequitors uttered by military and industrial planners, these are not "tactical" weapons to be used for short-term political objectives, but a means of destroying the earth. The deployment of Pershing and Cruise missiles in western Europe going on today will reduce the time between launch and impact considerably. A nuclear war could begin through simple miscalculation. By shortening the time missiles are in the air we simultaneously push our adversary against the wall. There are numerous well-documented instances of computer failures, human errors, and acts of God, that have sent our (and presumably the Soviet's) retaliatory mechanisms into motion. Thus far, there has been enough time to verify whatever it is that has caused the false alarm. As we reduce that time interval, we make it more probable that a miscalculation will result in cataclysm. This has resulted in what is called *Launch on Warning*, a condition where rather than waiting for further verification that an actual attack is taking place, the full retaliatory capability of the two superpowers will be on hair-trigger alert, and will be employed at the first sign of hostility.

Living under the constant threat of nuclear holocaust has exerted a profound influence upon the thinking and feelings of people, especially in the highly developed northern hemisphere. Psychological malaise is settling into the minds of young and old alike. It is often expressed in statements such as, "I do not wish to be a survivor." Given the level of ignorance concerning thermonuclear war and its aftermath, this attitude is not surprising. What must be kept in mind is that there were hundreds of thousands of survivors of the Hiroshima and Nagasaki bombings who were within a few kilometers of the blasts, and that they were abysmally ignorant of procedures to help themselves. So, while the governments of the superpowers continue to improve upon already unimaginable levels of destructive power, we, the common citizenry, must vanquish our pessimism in favor of knowledge, and preparation. The possibility exists that even in the event of a nuclear catastrophe all may not be lost and we may be among the survivors.

2. CRISIS RELOCATION PLAN: FACADE OR FACT?

In the early 1950s several government agencies were involved in planning to protect the civilian population during nuclear war. The objective of the Office of Civil Defense "was to help state and local governments provide a means for saving lives in the event of nuclear attack."[1] Beginning in 1961, buildings all over the United States were surveyed to determine if they were suitable for use as shelters. This task was called the National Fallout Shelter Survey, and was conducted by the U.S. Army Corps of Engineers and the U.S. Naval Facilities and Engineering Command. Buildings designated "suitable" were stocked with water barrels (often never filled), food, medical supplies, and radiation detection equipment. After several years of storage most of these supplies fell victim to pilfering, neglect, outright theft, and deterioration.

Concurrently, community warning systems were developed using existing town sirens and a commercial radio alert system. Millions of Civil Defense manuals were distributed containing information on personal and family survival which encouraged storage of supplies in private homes and the building of fallout shelters. Medical Self-Help Courses were promoted, but were generally poorly attended. These efforts continued in one form or another through 1966. After that, the popularity of Civil Defense waned and the multi-billion dollar program became history. Though it was a serious attempt to protect the civilian population against the effects of nuclear war, it simply did not catch on with the public.

Reports of a well-coordinated and highly developed system of civil defense in the Soviet Union in the late 1970s again stirred the interest of public figures. The Federal Emergency Management Agency (FEMA), which replaced the Office of Civil Defense, has developed a program of mass evacuation called the Crisis Relocation Plan (CRP). The agency claims that it could ensure the survival of eighty percent of the U.S. population. It consists of the planned evacuation of cities thought to be targeted to designated host areas, and assumes several days in which its plans could be implemented. It has been called by Howard Kornfeld in the *Western Journal of Medicine*, "A master plan indeed fraught with problems, and destined to failure if it is ever actually used in a crisis."[2]

Among the problems are the scale of the program itself, public ignorance of it, the inability to "test" it, the predictable public panic and unpredictable consequences of that panic, separation of family members, a short supply of food and other essentials, the breakdown of communications that would result from abandoning urban areas, the overcrowded and inhospitable host areas, and on and on. Can our government be serious in proposing such a plan, or is it, as Herman Kahn noted, simply "going through the motions of doing 'something'" so as not to "say in effect, 'We can no longer protect you in a war'?"[3]

Any survey of the evidence indicates that such programs are placebic in nature, and that survivability will only be likely on the most local of levels. Who would choose to separate themselves from their families when the threat of thermonuclear war is present? Who would simply wait his turn to be evacuated? What professional groups (medical, military) would remain intact to carry out such an incredible undertaking in a real crisis? Finally, the facts of nuclear warfare fatally undermine the idea of mass evacuation. Strategic missiles in the Soviet Union can reach their targets in the U.S. in thirty minutes; those based on submarines in ten minutes or less. The old system of the Office of Civil Defense was at least realistic in acknowledging that any program that is intended to work, and not just salve the consciences of congressmen, must be based on the smallest community level possible. People living in targeted areas are generally aware of that fact. If the time needed to carry out the CRP is available, the freeways will be clogged, public transportation will be overwhelmed, and anyone capable of functioning as a host will have less than generous thoughts on his mind.

3. EMERGENCY MEDICAL CARE: ILLUSION OR REALITY?

SINCE the Civil War our nation has not had battles fought within its borders. This has resulted from our geographical location and our relative power compared to the rest of the western hemisphere. But this isolation has become irrelevant in this day of thermonuclear war where entire continents can become the fields of battle. In this century, our society has evolved from being predominantly agricultural to being primarily industrial, and more recently, service oriented. Because people tend to live near where they work, our population has become concentrated into urban areas, and has also dramatically increased. Urbanization significantly increases the effectiveness of nuclear weapons, because it reduces the number of targets and increases their size and importance. (In contrast, the Soviet Union is still predominantly rural and agricultural, the one best, unstated reason that civil defense is a more realizable goal there.) Most strategically important industries are located close to population centers and certainly will be top priorities for Soviet planners. Therefore, even if they decide to leave our cities alone and destroy only strategic targets (which would otherwise be used in retaliation against them), city dwellers will die in great numbers.

Obviously, if one lives in the center of a target area (urban or strategic) there is nothing that can be done in case of detonation. But many living on target fringe and rural areas can be expected to survive the actual blasts. Countless scenarios have been constructed of the results of such a war. The estimates of fatalities range from 500,000 to 200,000,000. Though the United States boasts the finest health care facilities and personnel in the world, most of them are centered in targeted urban areas, and therefore will be destroyed. Few survivors will have access to professional medical treatment. The dangers of radiation, fires, and chaos will be too great for those able to treat the injured to take the risk. Any medical care facility still operating will quickly be overflowing with casualties. Supplies will soon be depleted, and their staffs stressed to exhaustion. For example, consider the situation described by John Hersey in his

EMERGENCY MEDICAL CARE   11

book, *Hiroshima*: "By nightfall, ten thousand victims of the explosion had invaded the Red Cross Hospital.... Patients were dying by the hundreds, but there was nobody to carry away the corpses... thousands of patients and hundreds of dead were in the yard and driveway."[1]

The problems in Hiroshima and Nagasaki, where only two cities were destroyed with low-yield weapons, compared with an all-out thermonuclear war, where hundreds of cities could be destroyed with high-yield weapons, should serve as a warning. According to a report to the Secretary-General of the United Nations by the cities of Hiroshima and Nagasaki in 1976, deaths following the bombings by the end of 1945 totaled over 210,000, out of a combined population estimated at 515,371 (that is almost 41%). That statistic is drawn from Eisei Ishikawa and David L. Swain's translation of a Japanese book entitled *Hiroshima and Nagasaki: The Physical, Medical, and Social Effects of the Atomic Bombings*. Therein the incredible destruction and short and long-term effects are analyzed. The book documents the complete elimination of health care delivery systems. Prior to the bombings, both cities had elaborately designed first aid programs and evacuation plans. It was wartime, and they were well-prepared for conventional attack. After the bombings, conditions were ripe for the outbreak of epidemics, since sewage and water systems were severely disrupted. But they were spared some of the worst results imaginable, because the rest of their country was still intact. In addition, they received aid from the American occupation forces. Thus, many of the injured were removed to other locations, and food and medical supplies were rushed in.[2]

In an all-out thermonuclear war survivors should not expect, nor will they likely receive, outside help. They will be forced to rely entirely upon their own resources and ingenuity. Residual radiation and blast damage may isolate individual survivors for long periods of time. In 1981, the American Medical Association's House of Delegates announced: "It is in the spirit of concern that the AMA Board of Trustees believes it is incumbent upon the Association to inform the President and Congress of the medical consequences of nuclear war and that no medical response is possible."[3] A short time later the American College of Physicians issued a "position paper" confirming that stance: "The American College of Physicians believes that there can be no adequate preparedness for the devastating consequences of nuclear war; prevention is the only reasonable medical response to the hazards posed by nuclear weapons."[4]

12 THE RELUCTANT SURVIVORS

The most serious threat to those who survive will be the short-term acute radiation effects. Unless radiation levels are known, a sheltered person may be completely immobilized by fear. Any pain or illness, or even slight nausea or diarrhea, will trigger fear of radiation sickness. Whether in large groups or in families, such reactions will be destructive of hope and the ability to go on.

The health and medical professions clearly realize that nuclear war is the greatest threat to survival since the epidemics of plague (Black Death) and smallpox in the Middle Ages. Surviving medical resources could not be expected to care for millions of burns, blast trauma, and radiation sickness cases. The initial injuries will largely be confined to target areas. It seems reasonable that there will be many millions uninjured during and immediately after such a holocaust, but that they will face a future of soon-to-arrive radioactive fallout of uncertain intensity and duration, millions of refugee casualties throughout the nation, a crippled government at all levels, uncertain food resources, epidemics of communicable diseases—in short, a shattered society. This shocking image appeared in the *New York Times*: "Imagine the familiar landscape turning suddenly into a sea of destruction: Everywhere smoldering banks of debris; everywhere the sights and sounds and smells of death. Imagine that other survivors are wandering about with bleeding wounds, broken limbs, and bodies so badly burned that their features appear to be melting and their flesh is peeling away in great raw folds."[5]

Further complicating treatment is the fact that "the initial symptoms of radiation sickness—namely vomiting and bloody diarrhea—are almost the same regardless of what dose of radiation has been absorbed .... It will be almost impossible to know for sure who might survive, if given medical support, and who will die no matter what is done for them."[6]

In almost all planning and disaster scenarios the assumption is made that physicians, nurses, and other medical professionals who are not killed or injured would go forth immediately and begin caring for the wounded. This implies that they will be immune to the psychological shock that other survivors would be suffering, would valiantly set forth despite high radiation levels, forgetting their concern for family and friends, and work totally without regard for self. Isn't it more logical to believe that they, being human like the rest, would seek shelter and family? And, as Christine Cassel expresses it: "If some physicians and nurses of extraordinary altruism, or simply obsessively patterned behavior, were to set

EMERGENCY MEDICAL CARE 13

about to care for the injured, what would be the objectives of treatment? What could be done?"[7]

The organization best suited for survival is small. It is a cohesive group which can meet crises under the most demanding conditions. It is the family. While it is true that there can be no adequate response by the health and medical community to thermonuclear war, a reasonably adequate response *is* possible at the family level, provided they:

1. Understand the dangers.
2. Prepare protective measures.
3. Become thoroughly familiar with certain selected health and medical information.
4. Implement preventive health measures before and during attack.
5. Provide treatment if injury occurs.

Clearly, persons receiving extensive thermal or blast injury, or who have absorbed very high levels of ionizing radiation (or combinations of these three effects) would not be expected to survive. But who will these be? Do you know exactly where and what you will be doing at the onset of a thermonuclear war? For those in cities which may not be hit, and for those in sparsely settled and urban fringe areas, the primary danger will be fallout. For those living in cities that will be hit, but who somehow survive—the information which follows provides you and your family a much better chance of limiting the damaging effects. These suggestions for enhancing the likelihood of survival are based on sound scientific research in nuclear physics, pharmacy, chemistry, medicine, nursing, public health, and radiobiology.

PART II

WHAT follows is an explanation of the dangers survivors will face after a thermonuclear war. First, we deal with the strange phenomenon of electromagnetic pulse (EMP), along with ways to protect your home electrical devices. Next, we discuss flash blindness, the result of exposure of the eyes to the brilliant flash of a nuclear detonation, and ways to treat it. We then move on to Black Rain, a commonly observed occurrence in many tests, that has its own peculiar dangers. At that point, we come to the real focus of the book, the prevention and treatment of radiation sickness. We discuss the nature of the various types of radiation, how it is distributed as fallout, and how to shield and otherwise protect yourself from it. The decontamination of food and water is then investigated, along with various techniques to protect goods beforehand. Finally, we discuss the nature of the malady itself, the various preventive measures that can minimize the risks of exposure, and explore the best commonly available treatments.

4. NUCLEAR PHENOMENA

ELECTROMAGNETIC PULSE (EMP)

SINCE 1945 we have been bombarded with descriptions of the terrifying destructive power of nuclear weapons. The effects of four of the ways that resultant energy from a nuclear fusion process will be distributed have been well-documented and publicized. These include (1) blast and shock waves, (2) thermal (heat) radiation, (3) initial irradiation, and (4) residual radiation. A fifth, though not nearly as well understood by the general public, may ultimately decide the "victor" in a future war, and be as significant as any of the others. Electromagnetic pulse (EMP) is similar to radio waves and occurs when high levels of gamma radiation are absorbed into the atmosphere. As the result of a thermonuclear detonation, the EMP can amount to thousands of volts being introduced into all receiving antennae within range of the waves. In effect, any electrical device that could possibly be "struck" could have thousands of volts running through it. Though it is a one-shot phenomenon, one dose would be more than enough to damage most conventional electrical systems. At the very least, most devices exposed would suffer burnt-out capacitors, transistors, and other components, thus requiring repair. Michael Riordan, in his book *The Day After Midnight*, points out that much sophisticated electronic hardware, especially that monitored by computer, cannot suffer temporary losses of power without requiring substantial effort to again start up.[1] The military implications are sobering.

During the time the "race track" basing mode for the MX was being promoted for the Great Basin of western Utah and eastern Nevada, the author had the opportunity to attend several briefings and hearings, where the public was allowed to ask questions of military planners, politicians, and scientists. During a lull in one of these, I asked several Air Force officers two questions: (1) "If the Soviets attempt to destroy the MX system by saturating the area with nuclear weapons, would the resulting intense radiation levels cause the MX warheads to detonate?" and (2) "If the Soviets try to destroy the MX system with nuclear weapons, wouldn't there be a tremendous amount of electromagnetic disturbance in the atmosphere in the vicinity of the system? If so, how could we launch a retaliatory strike?" My first question elicited polite laughter, and a reassuring, paternal, "now, my friend, you don't need to worry about high radiation levels detonating our warheads." That possibility, however, is a common subject of speculation in the literature of the arms race. In response to my second question one asked sharply and with some trepidation, "Where did you hear about that?" Two others walked away. Sensing I had stumbled onto something interesting, I hammered away. Again, I asked, "Could we launch an ICBM through the intense radiation and electromagnetic fields brought about by a Soviet strike?" I was left with one officer, a retired Air Force general, who I determined was concerned about this problem. He reluctantly answered my question. "We don't like to discuss this particular aspect. We do not really know. There is the possibility that we couldn't [launch]." His voice trailed off, filled with apprehension.

That took place in 1980. Since then, much new material on the subject has appeared. Herman Kahn was one of the first to write about it in the 1950s in his book, *On Thermonuclear War*, but he did not use the term "electromagnetic pulse." He called it "electromagnetic radiation" in listing it as one of the effects of a detonation. Later in the same book he discussed the blacking out of communications over a 3000 square mile area in the Pacific after an atmospheric test.[2] Clearly, any communications network or facility unprepared for the phenomenon could be rendered useless. Since our entire military apparatus depends upon electricity and communications, both of which would be shut down, our defense planners have been intensely concerned with this. This concern is reflected in the debates concerning the so-called "hardening" of defense facilities and missile silos.[3] One reassuring note, however. Our most reliable means of deterrence remains our fleet of nuclear submarines, which would remain unaffected.

But how will it affect your family and mine? A sudden, total loss of electronic communications would unnerve the public. Unprotected TV, teletype, radio stations, and telephone systems would be quieted, thereby vastly complicating emergency measures (such as evacuation) and community organization. Conceivably, air raid sirens could be knocked out. The circuitry in automobiles might be burned. The scenes in *The Day After* concerning the impact of the phenomenon on the public is evidence of how it has been acknowledged as a major problem. What in our urban society does not depend upon electricity? Water delivery, sewage treatment, etc., etc., ...

Dealing with these problems is beyond the range of this book. The well-prepared individual can only try to minimize the damage to his home, which will seem a significant achievement if he finds himself after a few days trying to listen to the *one* radio station that has gotten back on the air.

Methods of Protection Against EMP

1. If warning of an imminent nuclear attack is given, immediately unplug household appliances, including radios, TV sets, and CB equipment.
2. Disconnect antenna from TV sets, CB equipment, radio receivers, and shortwave communications equipment.
3. Remove automobile batteries and take into your shelter area as a power source. Because of the short length of auto wires there is a possibility auto systems may not be affected by EMP.
4. Take battery powered radio receivers and batteries to the area of the house or shelter in which you and your family have planned to stay during nuclear attack.
5. AM radio receivers have built-in ferrite or loop antennas. They will not be affected by EMP. AM/FM radios, in addition to the loop antenna contain extendable antennas for FM reception. DO NOT EXTEND THE FM ANTENNA. By not extending the FM antenna, the radio is protected. Placing radio receivers in a metal box will afford added protection against EMP (without long wire antennas connected).
6. CB, AM, and FM radio antennas less than twelve inches long will probably not be affected by EMP.
7. Ground wires should be removed from appliances (until EMP dissipates), and CB, FM, and AM radio equipment should not be grounded.
8. Unshielded radio equipment should be kept several feet away from sheet metal, pipes, or wires, since these can conduct high voltage surges of EMP.
9. Mobile CB, FM, AM, and shortwave transceivers are subject to EMP. The electrical power sources for these radios should be disconnected, and antennas removed. If not removable, the antenna connection should be grounded to the vehicle. Wrapping the unplugged unit in aluminum foil may offer it additional protection.
10. Heat pumps, electric fan motors and electric pump motors are

subject to EMP. Disconnection may afford adequate protection.

11. Flashlights and candles may be used in crowded shelter areas when the electric power is out. Kerosene, propane, and butane heaters are dangerous not only because of fire hazards, but also because they rapidly deplete oxygen in closed quarters. High concentrations of carbon monoxide and carbon dioxide gas can cause illness and death.

12. Low current-drain, battery operated lights are now available which are easily installed in closets, stairwells, or shelters. They are inexpensive and operate on C or D cells. Extra bulbs for these units should be included in preparations. These battery operated light sources are not affected by EMP.

FLASH BLINDNESS

The first perceptible indication of a thermonuclear explosion is the flash. There are two rapid pulses. The first lasts for a second or two, and is comprised primarily of ultraviolet light which travels only short distances in the atmosphere. The second arrives immediately after, and at peak intensity may be 100 times brighter than the sun at noon on a clear day. It is comprised of infra-red rays and light. The flash will last from approximately 22 seconds (one megaton burst) to slightly over a minute and a half (twenty megaton burst).

Observers of aboveground nuclear tests were provided special goggles or glasses of "smoked glass" and were told to use them when looking at the explosion. Lesions of the cornea, conjunctiva, and retina can be caused by the flash effect even when people are far enough away from the explosion not to be affected by heat, blast, or ionizing radiation. There are also ocular lesions which result from neutron exposure and which accompany radiation sickness. These will be discussed in a later chapter.

Like almost every aspect of the nuclear phenomenon, the risk of flash blindness is minimized through shielding techniques. The greatest variable is the distance from the detonation. Also important are atmospheric conditions (weather), terrain (ridges or mountains between you and the fireball), time of day (the pupils are dilated at night, allowing in more light). Clearly, if the opportunity exists to be sheltered during detonation, then the problem is dissipated before one ever emerges. It is fundamental to understand that even this intense light can be more or less protected against by closing the eyes, averting one's gaze, and seeking cover behind anything that throws a shadow. Windows in an emergency shelter area should be shrouded well in advance of blasts. Most cases of flash blindness that have been documented, especially in Hiroshima and Nagasaki, were quite temporary. Many had their sight return within two to three minutes. This condition seems associated with victims viewing light from a blast that has been scattered, rather than from looking at the fireball itself. Direct viewing, even from great distances, can result in retinal burns which are permanent, and potentially very serious.[4]

Emergency Treatment—Flash Blindness

Never rub your eyes for this may cause infection or damage to the delicate covering of the eyeball. Close the eyes so that tears will accumulate, which may wash out foreign objects and prevent infection. Tears naturally contain lysozyme, which is bactericidal. A weak solution of salt water (½ teaspoon of salt per pint of water) may safely be used as an eye wash. A weak solution of boric acid may also be used. Cold, wet compresses applied lightly over the eyes may provide some relief, and help the eyes obtain needed rest. Eye injuries are agitating and can be most uncomfortable. Adults may require two aspirin tablets (5 grains each) and children one, as per package instructions. If serious injury is suspected, try to obtain medical assistance as soon as possible.

BLACK RAIN

In a nuclear detonation at the earth's surface or under water, immense amounts of material are sucked up into the resulting plume, forming what is commonly referred to as the mushroom shaped cloud. In part because of the great weight, the material does not reach high altitudes, and therefore does not spread as far as after an air burst. Following the bombings in Japan, this highly radioactive cloud very quickly began to "rain." Some people thought that the Americans were dropping gasoline, intending to burn them. The truth was in fact worse. The radioactive rain, debris, and dust was carried over populated areas not directly affected by the bombs.

The special consideration this aspect of nuclear detonations demands of potential victims is an awareness of how quickly the problems of radioactive contamination *can* occur. Though it does not always happen, when it does "black rain" is the first stage of fallout. During the hydrogen bomb tests at Bikini atoll, the first radioactive rain (formed from ocean water) reached the ground surface one minute after detonation. The factors that influence it and how quickly it begins are related to geography and the type of detonation (surface, air, water).[5]

5. FALLOUT AND NUCLEAR RADIATION

THE detonation of a thermonuclear weapon results in five major effects: (1) blast, (2) heat, (3) initial radiation, (4) residual radiation, and (5) electromagnetic pulse. The first three can cause serious damage over large areas compared with conventional weapons, but their effects are seen as quite limited when compared to the area endangered through fallout. Electromagnetic pulse has already been shown to be a threat to very large areas.

Wind patterns vary with the seasons. Thus, in the event of a nuclear attack, fallout distribution will differ according to the time of year. Rate of drift will depend upon wind speed; radioactive intensity upon the design of the bomb and whether it is detonated in the air, on the ground, or under water. The rate at which fallout is deposited will depend upon several factors. Among them are:

1. Distance from the point of detonation.
2. Type of nuclear weapon.
3. Size (yield) of weapon.
4. Prevailing wind speed and direction.
5. Surface and upper atmospheric weather conditions (rain, snow, temperature, etc.)

With so many variables to consider and the inherent difficulty of quantifying each at the actual time of a detonation (except under "test" conditions), it's not surprising that predictions are practically worthless. Aboveground tests in Nevada have shown that under the best conditions wind speed and direction can shift abruptly, and that fallout is rarely distributed evenly over an area in ever-diminishing amounts. It has been found that "hot spots" can develop at unpredictable distances from the point of detonation and that characteristics of terrain can have substantial impact on the fallout patterns.

The fission-fusion process which occurs when a thermonuclear weapon is detonated produces about 200 distinguishable radioactive isotopes. An isotope is a form of an element with similar chemical properties to its naturally occurring type, but of a different atomic weight. A "radioactive isotope" is an unstable form of such an element. Many have very short half lives, others emit radiation over a period of days or weeks, and a few are very long lived. Upon detonation of a one megaton thermonuclear weapon, there is a brief, extremely intense, acute flash of alpha, beta, and gamma rays, along with streams of freed neutrons, for the first twenty seconds. This results from the incredibly rapid decay of many of the radioactive isotopes formed by the detonation. At the same time, the air becomes heated to incandescence. After a few millionths of a second there appears a ball of fire which increases in size and becomes correspondingly brighter.

A ground burst (necessary against such "hard" targets as missile silos) sucks millions of tons of soil, rock, and other debris high into the atmosphere. Shortly thereafter, this material begins to fall back to earth. Often, this is accompanied by rain. Radiation has been induced into many of these particles. Many particles are so fine that they will be carried into the stratosphere to drift around the earth for many years, all the while decaying at a constant rate. The heavier, pea-sized materials will begin to fall a few miles from the point of detonation. These will be highly radioactive for a time. As the thermonuclear cloud dissipates, smaller particles will fall. Particulate matter in a radioactive cloud is thus transferred to ground surfaces, people, buildings, and vegetation by "fallout," "washout," and "rainout." It is customary to use the word "fallout" for all the processes by which surface areas become contaminated. The word, "fallout," is also used to denote the particulate matter deposited. Because of the number of variables, the extent and nature of fallout can range between wide extremes. It is important to remember that fallout is a gradual phenomenon, and a relative one, extending beyond the point the radioactive cloud is no longer visible. Although there are so many variables that equipment is necessary to measure the amount of radioactivity, there is a general rule to estimate the decrease in radioactivity over time. It is called the seven-fold rule, and means that generally speaking, for every sevenfold measure of elapsed time, the radioactive contamination level is $1/10$ as strong.[1] For example, the radiation level after 7 hours would be only $1/10$ what it was immediately after detonation of a bomb. After 49 hours (7 x 7) it would be only $1/100$ the original level. It is important to remember, however, that the original contamination level is the most important factor in judging one's safety over time. And, unfortunately that cannot be known without detection equipment. If that is not available, it is best to remain sheltered as long as possible or until notified it is safe to emerge.

An excellent example which demonstrates the distribution of fallout is the eruption of Mount St. Helens in the state of Washington May 18, 1980. It was estimated that the force of eruption was roughly equivalent to a thermonuclear bomb with a yield of fifteen megatons. Tremendous accumulations of ash fell in close proximity to the volcano in a very short time. As the cloud traveled eastward it fanned out, with lesser quantities falling in more or less direct proportion to the time elapsed and the distance from the volcano, until a large percentage of the country was receiving volcanic fallout. But there are two major differences between the Mount St. Helens eruption and the detonation of a fifteen megaton blast. First, the volcanic ash was not radioactive. Second, the volcano emitted ash over a much longer period of time than would a thermonuclear blast.

Radiation is a strange phenomenon: it can be deadly and yet cannot be seen, heard, smelled, felt or tasted. To begin to understand what radiation is we must categorize its various types, their major characteristics and precautions that can be effective against them. We will then examine ways radiation can enter or affect the body.

Types of Radiation

1. *Alpha particles* are positively charged particles emitted by certain radioactive isotopes. Alpha particles have very little penetrating power, and can be stopped by a sheet of paper. But when an alpha particle emitter is ingested, they can become a dangerous source of internal radiation.

2. *Beta particles* are electrons (negatively charged) and are considerably more penetrative than alpha particles because they travel at higher velocities. Beta particles can penetrate the skin to a depth of several millimeters, yet can be stopped by a sheet of tin or a thin sheet of aluminum. Fallout containing beta-emitting materials can cause radiation damage to the skin. If ingested by breathing, eating, or drinking, these particles can cause serious damage to body tissues.

3. *Gamma rays* are high energy X-rays, and are highly penetrative and destructive to tissue. Gamma radiation is very dangerous because it causes massive cell death, and is capable of penetrating most materials, even relatively thick sheets of lead.

4. *Neutrons* are sub-atomic particles which carry no electrical charge. They are equal in mass to the proton. Neutrons are highly penetrative and can cause even greater tissue damage (per unit) than gamma radiation. Neutrons have a high affinity for water, and

Fallout and Nuclear Radiation 23

body tissue is composed of 96 to 98 percent water. The so-called "neutron bomb" relies upon this characteristic, and is thus more destructive of life than of buildings, for example. However, after detonation of a thermonuclear weapon, the danger from neutron radiation is over after the first 20 seconds. Thus, fallout does not carry danger from neutrons.

Radiation From Materials Deposited Internally

Inhalation and ingestion are the principal means of entry for radioactive materials into the body. As long as any remains, the body is subjected to internal irradiation. There is an important difference between external and internal irradiation. External irradiation danger ends soon after the radioactive cloud has passed, or the individual has moved out of the contaminated area. Radiation injury that has already occurred may later develop into radiation sickness, even though there is no additional exposure. But internally deposited material continues to radiate, and the body's accumulated dose continues to increase, until the radioactive materials are excreted or radioactive decay renders the material innocuous. Internally emitting radioactive materials may therefore cause greater damage to individual body organs than external gamma and neutron radiation doses.

Inhalation of radioactive materials occurs when the individual is exposed to "radioactive aerosol"—the suspension of tiny radioactive particles in a gas or smoke—and it is breathed into the lungs. In a report to the National Committee on Radiation Protection and Measurements, it was asserted that approximately 25 percent of inhaled material is retained in the lungs; the remainder is breathed or spit out. Soluble material is absorbed by the lungs and bowel, while insoluble particles are eliminated slowly. The same report points out that ingestion of radioactive materials results primarily from drinking contaminated water or eating contaminated food. First, the mucosal membranes are irradiated; then the stomach and other organs are subjected to the radioactivity.[2]

In addition to the various mechanisms the body employs to rid itself of alien particles (breathing, coughing, spitting, mucosal discharge, excretion, urination) there is another factor reducing the danger of internal emitters: radioactive decay. In fact, radioactive decay is the process of giving off radiation by which radioactive materials (such as uranium) become stable (no longer radioactive). The emission of alpha, beta, and gamma rays is what can be detected by a Geiger counter. The rate of decay of a radionuclide (radioactive substance) is called its "half life," which means that half of its radioactive energy has been given off during that span. We must remember that half still remains. Thus, for example, if a radioactive isotope had a half life of 8 days, it would have half its original radioactive energy after 8 days. In another 8 days it would dissipate half of the remaining energy—thus leaving it with one-quarter of the original amount; and so on, until the energy remaining would be so small as to be unmeasurable.

Protection Against Radioactive Fallout

People living in large urban areas should take the same precautions as those living in small towns and rural areas. Be as prepared as possible at all times—but especially during times of serious political crisis. In areas reasonably far removed from prime nuclear targets, after the detonation of a thermonuclear weapon, there may be anywhere from a few hours to a few days before radioactive fallout occurs. Even so, do not plan to shop at commercial stores to obtain necessary supplies that should have been purchased long before. Stores will be crowded with unprepared individuals, and you would probably not find what you needed anyway. The interval between attack and fallout must be used to greatest advantage to prepare you and your family for survival.

A basement offers fairly good protection, provided you have made preparations in advance, such as the storage of water, food, clothing, medications, bedding, and other supplies. Generally, the basement area offering the greatest protection from radioactive fallout is the center space, away from exterior walls and windows. Wisdom dictates that you ready an area in your basement NOW, in case you and your family need it. If a thermonuclear war begins, you will not have sufficient time to BEGIN making ready. That will be the time to make a few final preparations. Hold a family meeting now, and tell members what needs to be done. Form a plan and follow it.

There are three ways in which the damaging effects of radiation are reduced: Time, Distance, and Shielding. Each of these factors works with the others, such that a balance of them affords what can be called "effective" protection. As time passes, the radioactivity decreases. It is therefore important to be able to limit exposure for as long as possible. This points up the necessity of supplying shelter areas with enough goods to see your family through the worst of the fallout periods. Unless you are equipped with radiation detection equipment, it is best to plan to stay indoors for two weeks after nuclear fallout begins. Even after that much time, residual radiation levels might force people to stay in shelters, except for limited periods.

Distance from the source of radiation is another important factor in determining risk. An individual on the ground amidst radioactive fallout would be exposed to substantially more danger than would a person in a tower twenty feet above him. In the same way that distance from the detonation reduces risk from blast and heat effects, distance from radiating materials reduces the likelihood of exposure. Therefore, locate your shelter as far from exterior walls and windows as possible. This is the reason that the intermediate floors of tall buildings can afford excellent fallout protection: The distance from the ground and from the roof, where fallout will accumulate, mitigates its likely impact on people.

Shielding against damaging rays is the other obvious means of protecting against radiation. Since some radiation is highly penetrative (gamma rays, for example), heavy, dense materials make for the best protection. Generally, the heavier the shield, the better the protection.

Suggested Activities Between the Initial Attack and Arrival of Fallout

It is suggested that the following preparations be made during the time between attack and the arrival of fallout. All family members should work efficiently, cooperatively, and with minimum delay. The family should be instructed in advance on each of these tasks, and provided with the reasons why each is necessary.

1. Ensure that previously stored water, food items, etc. are intact, and easily accessible. Additional water and food should be added from other supplies at this time. Unharvested food, if available, should be gathered after other last minute preparations are completed, if time permits. The types of foods and quantities stored should have been determined and purchased previously. Many publications are available on the subject.

2. The minimum adult requirement of water is 2 quarts per day, which includes cooking and drinking. However, forethought indicates that at least one gallon per person per day is more realistic. Water for sponge baths and other uses should be available, and will make life much more endurable. In communities located distant from prime target areas it is possible that there will be no interruption of water, sewer, and electrical services during this period; but don't count on it. If water is available from the community system, it may become contaminated with radioactive fallout if the source

is a surface supply: (especially river, reservoir, or other open source). However, if the municipal system derives its water from an underground well, or springs, or a combination of both, in all probability the water will not become contaminated by radiation. EMP surges may render electrically-powered pumps inoperative, however.

3. Toilet articles and sanitary supplies should be provided in sufficient quantity for at least 2 weeks.

4. Enough bedding and clothing should be available. Without heat, even in summer months, basements can become cold. During early spring, late fall, and winter, prepare for the worst possible weather.

5. Individual medications and a well-stocked first aid kit will be crucial. SPECIFIC ITEMS FOR PREVENTION AND TREATMENT OF RADIATION SICKNESS OUTLINED IN PART II, CHAPTERS 8 AND 9 ARE ESSENTIAL.

6. If your home is supplied with a wood and coal burning stove or a fireplace, even if not in the basement, you are fortunate. Coal, wood, and fire-starting materials should be stockpiled in as large quantities as possible without creating a fire hazard. As radiation levels decrease to a point where short exposures are possible, fires for warmth and cooking can be built if the stove is located on the ground floor. If it is located in the basement it should be used very sparingly, if at all, when radiation intensity is extremely high, because of the danger of depleting the oxygen supply in closed and crowded quarters, and because the draft required might draw in radioactive dust from outside. In lieu of a stove or fireplace, two five-gallon propane bottles with valves and regulators in combination with a one-burner or two-burner camp type stove may be used as little as possible for cooking and heating.

7. Using wide plastic or masking tape, all windows should be shut, locked, and taped around the areas where radioactive dust may enter. Doors to the outside should be taped around the edges and frames, on the tops and sides. The bottoms of doors should have rugs, old blankets, or sheets placed along them to filter out radioactive dust, and yet allow some passage of air into the house. Exterior air vents, including the electric dryer vent, should be taped closed to prevent radioactive dust from entering. Any spaces between the top of the foundation and the upper portion of the house should be stuffed with insulation, caulking, rags, paper, etc.

8. Farm animals and domestic pets should be moved into barns or sheds or other suitable buildings, provided with feed for a minimum of 4 days and plenty of water. It is not recommended that

household pets share the same areas of the home with family members during crisis periods.

9. Other essential items are flashlights (with extra batteries and bulbs), battery operated radios, preferably tube types rather than transistor types, and extra batteries. Tube type radio receivers are very resistant to EMP effects. A radiation detection meter is a high priority item, although quite expensive. The availability of a radiation meter will provide you with vital information on radiation levels both inside and outside the home.

10. Entertainment materials will help to pass many hours. They also tend to ease stress and worry. Books, games, hobbies, etc. are a few possibilities. Writing materials for recording experiences and keeping track of food, water, and other supplies will be useful. If illness occurs, recording the time of onset, symptoms present, fever, medications administered, etc., will be very useful if prolonged treatment is required. Detailed and accurate notes of the course of an illness may be very valuable in determining the outcome, especially if the illness is radiation sickness.

11. Automobiles, machinery, bicycles, motorcycles, etc. should be placed inside garages or other buildings, if possible. Otherwise, they should be covered with plastic sheets or tarps. Decontamination of these items will be much easier later if they are covered or inside.

These precautions should be treated as a checklist. There are many books available that thoroughly discuss preparedness for nuclear war, and treat each aspect of life in a shelter comprehensively. It is highly recommended that all concerned persons review the literature so as to be as well-equipped as possible. Cresson H. Kearny's *Nuclear War Survival Skills* is among the most complete and usable of these books. (See Bibliography).

6. PROTECTION OF FOOD AND WATER

An all-out thermonuclear war can be expected to destroy only a small part of the agricultural land of the U.S. because of blast and heat. It will, however, likely contaminate a great deal more with varying levels of radioactive fallout, in part because many top strategic targets are located upwind of major U.S. farming areas. Fallout can affect agriculture in many ways. The first is by preventing workers from planting, tending, and harvesting existing crops. Second, it will likely kill and contaminate farm animals. Third, it can render crops on the vine useless and make land uncultivatable.

Much depends on the time of year. Since the ability to collect crops in the field will be limited by the same considerations that will keep people indoors, any use of existing crops will depend on their stage of development, the effects of untended exposure, the types of crops and their relative vulnerability to irradiation. Clearly, consumption after exposure requires thorough cleaning (decontamination). After the eruption of Mt. St. Helens substantial crop losses were suffered due to the shrouding of the plants under thick layers of volcanic ash. This prevented maturation. Some plants actually take up radioactive elements into their fruit, again depending on their stage of development. This subject is not as well-documented as others, though the susceptibility of certain crops has been quantified. Also, young plants are generally less resistant to radiation and are more likely to take up radioactive isotopes. Michael Riordan, in his book, *The Day After Midnight*, deals with various plants, and the likely effects, in some detail.[1]

The point to be remembered is that preparedness must include acceptance of the notion that agricultural productivity may be severely hampered, if not eliminated, in downwind areas, for a period of time. In most cases, by allowing sufficient time for the fallout to decay, and by plowing the radioactive material under, damaged fields can be restored to productivity. It is therefore imperative that dependence on agriculture be minimized as long as possible through adequate food storage.

Food Protection and Decontamination

Properly stored foods should remain free of contamination. This includes canned, bottled, and other foods stored in dust free, impervious vessels. According to the USDA 1966 *Yearbook of Agriculture,* "The principle of protecting food, feed, and water from external fallout is simple: prevent the fallout from becoming mixed or incorporated into these materials. They may be irradiated by the fallout, but if the radioactive particles do not come in actual contact with them—or if the fallout is removed—they will not be radioactive, and thus will be safe to eat or drink."[2] In other words, it is not exposure to radiation that makes food unusable, for it does not as a result become radioactive. It is only if food or water contains radioactive material that it becomes dangerous as a cause of internal radiation. Thus, precautions must be taken.

The family interested in minimizing vulnerability should have a food storage area within their designated shelter or basement that will not be subject to contamination. The methods to employ are the same as those used to keep things dust free. Similarly, if foodstuffs have become covered with radioactive particles, washing, vacuum cleaning, etc., should remove them. Of course, caution must be exercised to avoid inhaling or swallowing radioactive material. It is to be hoped that within a short period of time (two to three weeks) some semblance of normalcy will have returned. This will largely depend on considerations specific to an area. If one is a survivor in a location spared all effects except residual radiation (fallout), the likelihood is that outside aid and community organization will soon provide verifiably uncontaminated food. If instead one lives close to a target area it would be prudent to be prepared for the worst. The Church of Jesus Christ of Latter-Day Saints (the Mormons) has instituted a family emergency plan that calls for at least one year of food storage. This well thought-out program is an excellent model. In any case, it should be noted that the United States maintains relatively small stockpiles of most essential goods, and that the distribution system would be largely destroyed in a thermonuclear war. Remember also that American agriculture is almost completely mechanized and that it cannot be successfully converted to nineteenth-century methods overnight. Scenarios that envision significant use of the products of commercial agriculture ignore the critical dependence of our economy on our ability to transport perishables great distances; and our transportation and fuel systems would probably be severely disrupted in the aftermath of a major nuclear exchange.

On the family level, then, it is necessary to understand what can be done with the produce of the backyard garden, and other contaminated foods that may somehow be obtained. The 1966 USDA *Yearbook of Agriculture* covers this subject quite thoroughly:

> Fallout particles that fall directly on food and forage plants contaminate them by remaining attached to the aboveground parts, or by releasing radio-isotopes that are absorbed into the leaves and other plant parts. Rain and wind move these particles from plants to soil, but certain characteristics of the leaves, such as hairiness, waxiness, and roughness, increase retention while smoothness reduces it.
>
> ... Accordingly, internal contamination ... from leaf absorption is greatest in leaves and is comparatively less in fruits, seeds, and edible roots and tubers.
>
> As the season progresses, the fallout contaminants that have accumulated in the leaves, ... may be washed from the leaves to the soil in rain and dew. They may then be absorbed by the roots and distributed throughout the entire plant and thereby increase the total content of contaminants.
>
> ... Vegetables and fruits harvested from fallout zones will require decontamination before they can be used for food. First, the exposed parts must be thoroughly washed to remove the fallout particles. Then vegetables or fruits should be peeled, pared, or the outside otherwise removed in such a way that hands or utensils do not contaminate the parts to be eaten. It should be possible to decontaminate almost completely such crops as apples, head lettuce, and cabbage by repeated parings, washing hands and washing utensils before each paring. Since fresh fallout provides only surface contamination, it should be possible to wash and shell peas and beans or to husk sweet corn in order to remove the contaminated parts.
>
> This type of decontamination could be applied to many human food items in the home immediately after harvest, if possible using well water, or some other noncontaminated water. It should be remembered, however, that one can wash his hands effectively using dirty water, and that it is also possible to decontaminate most vegetables effectively using radio-contaminated water for washing. Do not use drinking water for decontamination unless you have more than is necessary for that purpose. Cooking will not destroy radioactivity, but research has shown that boiling foods, including meat, will leach radionuclides from the food. The food itself may be consumed, but not the water it is cooked with.
>
> Some food products that have fallout on or mixed in them can be used only after holding the products long enough to allow the radioactivity to decay to a safe level. Storage of the contaminated material for a period ranging from 2 weeks to many months,

depending upon the degree and kind of contamination, will reduce the amount of radioactivity present, usually to a negligible level. Obviously, many food products—including most meat that is not canned—could not be stored for the necessary time....

With the breakdown of refrigeration, which is very likely in a damaged area, it may be impossible to salvage perishable products. But if spoilage is not too great, the perishables may be washed or trimmed and cooked thoroughly before eating them.[3]

The prospect of searching through the rubble of one's city for uncontaminated food is not pleasant. But if it comes to that, it is necessary to know some procedures for distinguishing safe from unsafe. A radiation detection device is the most trustworthy tool. Short of that, examine carefully anything that may have been exposed to drifting radioactive particles, keeping in mind that they can enter anywhere the finest dust can. Boxed and sacked goods are comparatively vulnerable because the containers are porous. Plastic bags, on the other hand, if properly sealed, are quite effective. Canned goods should be usable, unless dented or rusted (either condition makes likely holes in the can). Questionable cans should be stored as long as possible. Swelling would indicate bacterial spoilage, and such cans would have to be thrown out. Bottles are less safe because the possibility of ruptured seals between lids and containers is greater, and results in the same condition. Caution must be emphasized in the handling and decontamination of any food, regardless of packaging.

Water and Water Supplies

Water may be contaminated in several ways, such as fallout dropping into a river or reservoir, accumulation of radioactive fallout particles on watersheds, explosion of a thermonuclear bomb in or near a reservoir, or the deliberate use of radio-isotopes in radiological warfare. If the degree of contamination is not too severe, then it is probable that as a result of several factors the water will not be rendered unfit for consumption, except for a limited time immediately following contamination.

In surface waters, radioactive contaminants will tend to be adsorbed by the suspended and colloidal matter that is invariably present. ("Adsorption" is the process of adhesion of molecules to a surface where there is no chemical bonding. It is the opposite of "absorption.") This matter will partly settle or be adsorbed by the walls and bottom of the reservoir. In urban water systems radioactive material escaping adsorption in the reservoir itself may be picked up by the surfaces of the distribution systems which usually consist of highly adsorbent brick or rusted iron. When, in addition, the purification process includes coagulation, sedimentation, and filtration stages, it is to be expected that very little radioactive material will reach the consumer.

Because of the adsorptive properties of soil, underground sources of water are generally safe from contamination. For the same reason, moderately deep wells, even under contaminated ground, can be used as sources of drinking water, provided surface drainage of contaminated material into the well is prevented.

If a reservoir or river is seriously contaminated, and the water is not subjected to coagulation or filtration, the water may be unfit for consumption for several days. However, because of dilution and natural decay of the radioactivity, the degree of contamination will decrease with time. It would be necessary, in cases of this kind, to subject the water to careful examination for radioactivity and to withhold the supply until it is reasonably safe for human consumption. It should be remembered in this connection that since water is taken internally, alpha and beta activity as well as gamma activity is important.

In the event a family must remain indoors for two weeks to one month because of fallout, municipal water systems should not be depended upon, because of the possibility of contamination, structural damage to the system, loss of electrical power for operations, or a combination of these problems.

Large bottles do not use as much space as small ones for equal quantities of water, but five-gallon bottles are heavy and therefore difficult to lift and handle. Gallon containers are much easier to use. The risk of spilling precious water or breaking the bottle (whether glass or plastic) increases as the weight of the container increases. One or two-gallon-sized plastic bottles with tight-fitting lids are best for home storage. Plastic containers may affect the taste of the water over several months of storage, but the change is harmless. There are 30 gallon plastic water containers available with spigots for home water storage. Canned water is also available which has a storage life of up to five years. Another method of storing water is in 20 gallon plastic garbage containers with the tops sealed around the edges with masking tape. Each 20 gallon plastic container holds one week's minimum supply for three people. Fruit jars, quart bottles and other types of glass containers may also be used. Metal containers are safe, but may impart tastes to the water.

It is not necessary to boil water before sealing it in containers, if it is from a known source. Stored water should be changed every three months. Many individuals have purchased water beds as a means of storing water. If this method is used, the distributor or manufacturer should be contacted regarding the use of purifiers, algicides, and other information relevant to water storage.

If, in an emergency, you find you have failed to change water at appropriate intervals and it has an off odor or taste, don't throw it out. It can most likely be used as is, and, if necessary, can be purified.

In most homes the hot water tank holds from 20 to 80 gallons that are perfectly safe to drink. But remember, the gas or electricity must be turned off before draining. If the system is still in operation and no alternative source is available, home water softeners might be used to remove, or substantially reduce, radioactivity in water. By running water through the appliance, significant decontamination should occur.

Water in the pipes of homes and buildings may be used safely, unless the system has already had water introduced which contains radioactive contamination. In emergencies, local authorities may instruct people to turn off the main water valves to homes and buildings to avoid having water drain away in case of a rupture and the resultant loss of pressure in water mains. If the main valve of a house or other building is closed, the pipes will still be full of water. To use this water, the faucet which is located at the highest point of the structure should be turned on (opened). This will let air into the system and allow water to drain without the interference of a vacuum in the lines when a faucet at the lowest level of the building is also opened.

Another source of water in homes is toilet tanks (not toilet bowls). Toilet tanks generally hold between five and ten gallons of water in reserve for flushing the bowl.

Some large buildings (hospitals, nursing homes, pharmacies, laboratories, etc.) may have small stills for distilling water. Some people are clever enough to rig up a simple still in their homes. Distilled water is generally safe from radioactive contaminants, but not from disease germs.

Canned juices and other foods high in water content will tend to reduce the use of stored water. Consumption of foods with high salt content causes people to drink more water. Therefore, such foods should be avoided if safe water is scarce.

If the only water available is contaminated with radioactive materials, do not throw it away. The radiation may decay rapidly, leaving the water far less dangerous. Larger radioactive particles tend to settle out if water is left undisturbed. A government study of the subject concludes "that in an emergency, water with many times the accepted tolerance limit could be used for drinking in limited amounts. Because of the rapid decay of fission products with time, the activity of the water and the corresponding hazard... would soon decrease."[4]

Although boiling or chemical treatment will destroy disease organisms they will not remove radioactive contaminants from water. The most efficient methods for removing radioactive contaminants from water are not available to most homeowners. However, two relatively effective methods are often accessible:

1. *Water Softeners*: Running water through a tank-type household water softener will remove a very high percentage of the suspended, and part of any dissolved, radioactive contaminants.

2. *Settling and Straining*: Much radioactive contamination may be removed from water by simply allowing water to settle, then straining it through a cloth, paper towels, or fine sand (6 inches deep).

Water Source Safety

Water from almost any source can be made safe from bacterial infestation by using one of the five techniques outlined later in this chapter. However, these methods do nothing about radioactive contamination. Radioactive materials should be removed from water before it is used for washing, cooking, or drinking. It is also important that auxiliary sources of water be located before stored supplies are depleted. Below is a list of potential sources listed in decreasing order of preference.

1. Water from deep wells and other totally enclosed systems, municipal or private.

2. Water from tightly covered municipal and rural reservoirs or water storage tanks, even if the main source is an open reservoir or stream.

3. Water from developed springs and artesian wells (provided it is not contaminated by surface runoff).

4. Shallow wells and infiltration galleries and pits. Water from these sources is usually safe, provided the systems are waterproof.

5. Water from deep lakes and reservoirs. Many radioactive isotopes are relatively water insoluble, and thus will fall to the bottom, rendering the surface water suitable for use.

6. Water from shallow lakes, ponds, and swamps. Much less safe because wind agitates shallow water more, causing debris to constantly resurface.

PROTECTION OF FOOD AND WATER 35

7. Water from moving sources, such as streams and rivers. The motion keeps the radioactive materials at the surface, and distributes it through the entire course. These sources are especially dangerous after snow melt or rain.
8. Roof collected rainwater collected in a cistern would be quite unsafe since it would probably contain high levels of contamination.
9. Water collected from snow pack that has had radionuclides deposited upon it.

Of equal if not greater danger than radioactive contamination of water supplies is the possibility of waterborne diseases. In the aftermath of a thermonuclear war, general sanitation and personal health practices are likely to break down. Many people in the surviving population will be weakened by exposure to ionizing radiation, inadequate nutrition, psychological stress, lack of rest, and other problems. Chlorine and other purifying agents may not be available. The death of humans and animals along with the presence of vermin (flies, cockroaches, rodents, lice, etc.) will likely result in epidemics of viral diseases. Even under optimal conditions there are many circumstances which lead to contamination of public and private water supplies with disease organisms. Waves of typhus, influenza, gastro-intestinal infections, and staph and strep infections can be expected to occur.

Purification Methods

These methods of purifying water are intended to destroy microorganisms capable of causing disease. If a supply of water has been stored too long, it may become cloudy, show signs of algae growth or have an unpleasant taste or odor. Although in all probability it is safe, the algae and odor might be most disagreeable. During a radiation emergency, the first water that should be used for drinking or food preparation is that which is known to be uncontaminated by radioactive materials and disease organisms. But in emergencies the quality of water is often unknown. There may, however, be one or several sources of water available which are "suspicious." Suspicious water includes that found in shallow wells, cisterns, broken water pipes, ponds, rivers, and lakes. It may also include stored water which has turned cloudy, or that tastes or smells strange. Purification will reduce odor and taste, and render the water in question safe. Muddy water, or water polluted with large amounts of vegetation or other organic matter, can be purified by several simple methods.

If the water in question has off tastes and odors, but is known to be safe, it may be made more palatable by adding pieces of wood charcoal, or charcoal briquets, shaking vigorously for a few seconds, then allowing to stand for a short time. Charcoal absorbs tastes and odors rapidly. Activated charcoal is especially effective for taste and odor removal.

A good general rule is: WHEN IN DOUBT, PURIFY. Straining water which is muddy or which contains large amounts of vegetable matter or insects (such as mosquito larvae or skates) through clean cloth or paper towels is the first step. But if straining or filtering is not possible, the water should be allowed to stand without being disturbed (preferably covered), so solids can settle to the bottom. Just because water is crystal clear does not mean that it is safe to drink.

WATER PURIFICATION

1. Boiling

If heat is available, water should be boiled for at least 15 minutes. To improve the taste of water after boiling, pour it back and forth from one clean container to another after it has cooled. This adds oxygen to the water. Boiling destroys disease organisms, but does not remove radioactive materials.

2. Water Purification

Water purification tablets which contain iodine or chlorine can be purchased at drugstores, sporting goods stores, chemical supply houses, etc. Follow label directions.

3. Iodination

Ordinary tincture of iodine may be used to purify small quantities of water. Add 8 drops of tincture of iodine to each gallon of *clear* water. Mix and let stand for 30 minutes before use. Add 16 drops of tincture of iodine to each gallon of *cloudy* water. Mix, and let stand for 30 minutes before use.

4. Chlorination

Water may be purified by adding chlorine to it in the form of a chlorine household bleach. Make sure the bleach is a "chlorine type." Check the label. Chlorine will be indicated by the term "calcium hypochlorite." Use 8 drops of chlorine bleach (full strength) for each gallon of *clear* water. Mix, and let stand for 30 minutes before use. Add 16 drops of chlorine bleach (full strength to each gallon of *cloudy* water. Mix, and let stand for 30 minutes before use.

5. HTH, or other 65% free available chlorine compound

HTH and other chlorine compounds may be purchased from swimming pool supply companies, chemical supply companies, or drugstores. These compounds are extremely corrosive, and must be handled with caution. They come in granular or powder form, and must be mixed with water to make a standard solution. DIRECTIONS: Dissolve 4 heaping tablespoons of powder or granules in one quart of water. Mix thoroughly, then use as you would chlorine bleach to purify water.

CAUTION: THESE METHODS OF WATER PURIFICATION DESTROY DISEASE CAUSING MICRO-ORGANISMS, BUT DO NOT REMOVE OR RENDER INNOCUOUS ANY RADIOACTIVE MATERIALS WHICH MAY BE PRESENT.

7. DECONTAMINATION

AFTER a thermonuclear explosion, it is impossible to predict how large the hazardous downwind area of high intensity radiation fallout will be or the shape it will take. It may extend from 50 to 600 miles or more, alternating hot spots with areas of little or no fallout. As noted above, many variables affect fallout patterns—weather conditions and terrain features being among the most important.

Anything—people, animals, homes, machinery—in a field of radioactive fallout is "literally immersed in a crossfire of radiation, consisting of a mixture of gamma and beta rays." Gamma radiation is the greatest hazard from fallout because of its high penetrability. Wind blowing radioactive dust into homes and shelters through cracks in doors, windows, ventilators, etc., can result in serious illness and death. Fission products adhering to the skin and clothing combine to irradiate the individual with gamma and beta radiation. Beta rays produce a much larger *skin* dose than gamma rays because gamma rays penetrate much more deeply into skin tissues. "The beta rays can produce a serious burn without gamma rays reaching a lethal dose."[1] Ordinary clothing provides some protection against beta radiation, but none against gamma radiation. If enough fallout remains on exposed skin surfaces, or filters through clothing, the beta radiation will result in skin burns which will look similar to severe thermal (heat) burns and will be very painful.

Significant amounts of radioactive material "falls out" in the vicinity of a blast soon after an explosion. People 70 miles away might have an hour or more in which to seek protection. One hundred miles downwind, fallout may not arrive for three hours or more, depending on wind speed and other variables. Information on large yield nuclear weapons (one megaton and above) is somewhat sketchy, but it appears that the major residual radiation threat does not occur at the site of detonation, but about 50 to 75 miles downwind. At these distances and beyond, the most dangerous radioactive isotopes contained in the fallout are iodine 131, iodine 133, strontium 89, strontium 90, barium 140, and cesium 137. (See appendix 1 for table of radioisotopes.)

Basically, there are three methods which may be used to minimize the hazards associated with radioactive contamination:
1. Burying contaminated materials in the ground or at sea.
2. Keeping the materials a safe distance from humans and animals for a sufficient period of time to allow the radioactivity to decay to a reasonably safe level.
3. Attempting to remove radioactive contaminants, that is, to decontaminate the material or structure.

The method used will depend on existing circumstances.

Except where radioactive solutions, such as radioactive rain or snow, soak into porous materials, like rope, textiles, unpainted or unvarnished wood, etc., or where neutrons have penetrated and induced activity to some depth, the decontamination will largely be restricted to surface areas. The problem of decontamination is thus, to a considerable degree, a problem of removing sufficient surface material to reduce the activity to the extent that it is no longer a hazard. The methods of surface removal may be divided into two main categories, chemical and physical. In the first, contamination is eliminated by making use of chemical reagents which, if sufficiently mild, will have a minor effect on the underlying material. In the second case, an appreciable thickness of the actual surface may be removed.

It should be understood that the radioactivity of radioisotopes is not changed in any way by chemical reactions. All that is accomplished is to convert the active isotope into a soluble compound, so that it can be detached and washed off in solution. Certain processes of decontamination, involving the use of detergents, represent a category between the chemical and physical.

Any decontamination process entails risk through exposure to radioactive materials. Therefore, a careful determination of the benefits must be made before attempting decontamination. An object or structure that is needed to help sustain life will be a high priority for decontamination. This could include automobiles, agricultural equipment, public health care facilities, airplanes, ships, and the parts of homes abandoned by families during the emergency. Except where a structure or object is composed of porous materials (unpainted and unvarnished wood, cloth, rope, brick, etc.) reasonably successful decontamination can usually be achieved. Depending upon the levels of radioactivity, individuals might be limited to very short periods in which to work to maintain personal safety.

Procedures for Inanimate Objects

After a period of fallout, homes, yards, cars, barns, sheds, etc., will be contaminated to varying degrees. When the radiation hazard has subsided to an acceptable level, hosing off these structures with water from a garden hose will remove a large percentage of the contaminants. This technique obviously depends on a functioning water delivery system. Care must be taken that inhalation or ingestion of radioactive contaminants does not occur. This may be accomplished by wearing a mask made of toweling or other material which will filter out dust particles. Eyes should be protected with tight-fitting goggles. A pair of rubber irrigation boots (or equal protection), worn in conjunction with a jump suit, will provide protection against skin burns.

From the book, *Effects of Nuclear Weapons*, comes these suggestions:

Household cleaning and scouring compounds, grease removers, detergents, paint cleaners, dry cleaning solvents, gasoline, etc., will all help to remove radioactive particles from surfaces. In carrying out these processes, care must be taken not to distribute the radioactivity. The cloths or other materials used should be buried and not burned, unless special incinerators, which prevent escape of the active materials, are available ....

Exposed surfaces especially in cities are usually covered with "industrial film" of grease and dirt. It has been observed, in many instances, that radioactive contamination attaches itself primarily to this film, and that its removal consequently affects considerable decontamination .... Removal of the industrial film, if present, may be the first step in decontamination. Besides producing some decrease in the radioactivity, the removal of the film will facilitate the action of the various agents used for more thorough decontamination....

Shielding, whether by distance, terrain, walls, structures, etc., must be used as advantageously as possible. For example, the decontamination of a building or a ship should be started from a suitable position in the interior, where the activity will probably be less than on the outside. In this connection it is recommended that installations of less strategic importance, in a situation where contamination is a possibility, should be provided with hosing down equipment controllable from the interior.

It is not possible to give a general rule concerning the areas from which decontamination should be instituted. In some cases it would be advantageous to start in a region where the activity is low, for this will not only make the operation less hazardous but will allow time for decay of the more highly contaminated portions. On the other hand, in certain circumstances it might be advisable first to carry out a quick, even if preliminary, decontamination of an area of high contamination in order to permit freedom of movement.[2]

Radiological monitoring is a most important protective measure during decontamination procedures. For the safety of individuals performing decontamination, the existing radiation levels must be known. The persons performing decontamination must be monitored as well as the materials being decontaminated. Radiological monitoring can properly and safely be performed only with appropriate instruments used by competently trained individuals. There should be no exceptions even under emergency conditions. Decontamination efforts at the family level should be delayed until monitoring can be performed, unless detection equipment has been obtained. Families should seriously consider the purchase of a radiation detection meter. In a fallout situation few things would take precedence as far as safety is concerned.

Ben Freedman, author of *The Sanitarian's Handbook*, deals exhaustively with decontamination of inanimate objects, structures, and materials. He cautions that: "A great amount of experimental work has been done to develop the most suitable decontamination agents and methods. Much useful information has been acquired but the easy method is still to be discovered."[3] The techniques he describes run the gamut from simple scrubbing with water (which, depending on the surface, can be as much as 30 to 50 percent effective) to planing, sanding, or otherwise physically removing contaminated sections. Among the best cleaning solutions are citric and other organic acids, non-corrosive inorganic acids, alkali solutions (especially for painted surfaces, because they remove contaminated particles along with the paint), and detergents (especially on greasy surfaces). The dangers inherent in the process of decontamination point up the necessity of properly storing items that will be needed during critical periods. The more tightly one's house is sealed, the simpler the subsequent job. With the other demanding problems that nuclear war will engender, adequate preparation in this area seems prudent.

In all decontamination work, the disposal of radioactive contaminants must be considered. Water used in decontamination should not be allowed to enter sources of culinary water. Combustible materials (paper, wood, pasteboard) should not be burned, which would once again radioactively recontaminate the area.

Under emergency conditions burial is probably the best method of disposal, since the general environment would be contaminated anyway. It is most important to not add to the burden of radiation to which people have been, or are being exposed. If soil, waterways, storm drains, and sanitary sewers are already contaminated, liquid flushings and drainings from decontamination efforts may be permitted to flow into the ground or into usual drainage channels.

Protecting Individuals

Radioactive contamination should be treated like any other poisonous substance. Radioactive materials should not come in contact with the hands or other parts of the body if at all possible. Careful personal cleanliness is recommended, although understandably difficult in emergency conditions.

Individuals performing decontamination work should be frequently monitored. At the end of a work period the hands should be washed thoroughly. At no time during the work should hand to nose or mouth occur. Radioactive materials may be absorbed into the body through cuts, particularly on the hands. No one performing decontamination work should smoke, chew gum, drink beverages or eat until thoroughly showering with soap and water, if at all possible. This will help protect against the deposition of radioactive material inside the body.

Again quoting the Los Alamos report:

The decontamination of personnel who have come into contact with radioactive material is, of course, a primary requirement. Normally, clothing will prevent access of the material to the skin. When contaminated, clothing should be removed and disposed of, by burial, for example, in such a manner as to prevent the spread of the radioactivity into uncontaminated areas, like the interior of buildings.

In a dire emergency, any clean uncontaminated material at hand, such as paper, straw, grass, leaves or sand, will remove activity from the skin, if applied vigorously. However, care must be taken not to tear the skin, or to drive the loosened material into wounds, body openings or skin folds.

If the soap and water treatment does not produce the desired decrease in activity, chemical agents, if available, may be used on the skin. Isotonic saline of pH 2, or dipilatory or keratolytic agents, such as a mixture of barium sulfide and starch will lead to the removal of the material held tenaciously by the skin. A dilute solution of sodium bicarbonate is useful, especially on mucous membranes ....

Radioactive substances ... when in close contact with the skin may represent a much greater hazard than would be indicated by an instrument held an inch or two away. Consequently, attention must be paid to cleansing any exposed surfaces of the body. A very fair degree of decontamination of the exposed skin can be achieved by vigorously rubbing with soap and water, paying particular attention to the hair, nails, skin folds, and areas surrounding body openings, and with due care to avoid abrasion. Certain synthetic detergents, of which many are on the market, e.g., soapless household cleansers, have been found to be especially effective in this connection.[4]

Hair

Hair tends to retain radioactive materials most tenaciously. If decontamination cannot be attained by thorough washing, the hair should be shaved from the head and body, and the scalp again washed with shampoo, soap, or detergent solution.

Showers

Several soap and water showers (not baths) will probably achieve decontamination of an individual.

Eyes

Since the eyes are very sensitive, if the possibility exists of radioactive contamination, weak solutions of either sodium bicarbonate or boric acid are the best for treatment. Apply either solution liberally with an eyedropper or eyewash cup.

Mucous Membranes

The use of dilute sodium bicarbonate solution for decontamination of the mouth, nose, and eyes is usually effective. Other body openings should be washed thoroughly with soap and water. Soap, which is a chemical combination of a fatty acid and an alkali, is a wetting, dispersing, and emulsifying agent and, therefore, a surface-active agent. Synthetic detergents consist of manufactured compounds not readily found in nature. A synthetic detergent can accomplish all that soaps can, and is more effective in high and low pH waters. Detergents, including soap, are not disinfectants except as they act mechanically to remove bacteria, dirt, oils, and radioactive particles.

Barium Sulfide and Starch

In very stubbornly held radioactive contaminants, a mixture of barium sulfide and starch with enough water added to form a thick paste may be used to decontaminate skin surfaces. This mixture has been found to be very effective.

Decontamination 43

Degreasers

Degreasers may be applied to the skin to remove radioactive contaminants provided that they are immediately flushed off with large quantities of water. Degreasers are irritating to the skin, but are effective in decontamination.

Strippable Coatings

As a last resort, special plastic and/or rubber compounds which dry into an elastic film upon exposure to air may be applied to skin surfaces, or the surfaces of inanimate materials, then peeled off. This method is expensive and is usually used only when contamination is severe. Self-adhering tapes (adhesive tape, masking tape, scotch tape, etc.) can decontaminate surfaces by being applied to the surface (skin, etc.) and then being stripped off, thus removing contaminants from the surface.

One last note of caution from the Los Alamos report: "In considering the problem of decontamination, there is one fundamental point which must not be forgotten. Decontamination procedures do not neutralize the radioactivity; they merely transfer the active material from one place to another. Consequently, before undertaking decontamination it is necessary to arrange for the proper disposal of the material removed, to a location where it does not represent a hazard. The method used must be determined by the circumstances existing at the time."[5]

During and immediately after a nuclear attack, there will be no organized neighborhood or community efforts at decontamination. At these times, the individual and family may be required to make decisions with respect to decontamination. A basic understanding of radiation, fallout, contamination, and their effects, combined with sound judgement may be the only basis upon which an emergency decision can be made. The practice of good, ordinary cleaning procedures should adequately decontaminate most household items. Some decontamination will take place naturally, as rain washes radioactive materials away, and as they leach into the soil. Time allows radiation levels to decay. But it is possible that some areas would have to be totally evacuated because of high radiation levels which persist long after initial radioactive contamination.

8. THE ACUTE RADIATION SYNDROME (RADIATION SICKNESS)

THE Acute Radiation Syndrome, or radiation sickness as it is more commonly known, is characterized by a variety of signs and symptoms. It is caused by exposure to ionizing radiation. "Ionizing radiation refers to electromagnetic waves, such as X-rays and gamma waves, and to particulate radiation such as neutrons and protons .... Ionizing radiation always produces some degree of damage to cells or tissues."[1] It does not matter whether the radiation comes from over-exposure to X-rays, direct nuclear radiation from the explosion of a nuclear weapon, or from fallout—the effects on the body are essentially the same.

There has been considerable misunderstanding resulting from dissemination of false information over the years about radiation sickness. The illness runs a peculiar yet predictable course. Much has been said and written by well-meaning but relatively uninformed individuals which has added to our fear about this strange malady. Most dependable material which describes radiation damage to the human body, symptoms, and eventual outcome of the illness, is explained in scientific terms and is usually incomprehensible to those not trained in medicine, biochemistry, or radiobiology. In this chapter an attempt is made to describe radiation sickness in simplified language, with a minimal use of technical terminology, so it can aid in better understanding this modern threat to mankind.

Acute Radiation Syndrome is the medical term for radiation sickness. The term *Acute* implies that the onset is sudden, as opposed to being a condition which develops over a longer period of time. The term *Syndrome* refers to a particular group of symptoms and signs which are characteristic of the illness. The exposure of humans to ionizing (nuclear) radiation results in a variety of symptoms—each of which is also symptomatic of other illnesses. When these symptoms occur after exposure to radiation, they indicate that certain physiological and biochemical systems of the body have been affected. Because the organ systems are adversely affected

RADIATION SICKNESS 45

at the same time it is unlikely that a single medicine, treatment, or other curative process will ever be found to treat all symptoms. This will be more fully understood as the damage to body tissues and develpment of the illness is explained in detail.

During the illness, patients do not become radioactive, nor does any part of the body become radioactive. Radiation is not induced into the body, even though radioactive materials can be inhaled or eaten. Radiation sickness is not communicable, that is, it cannot be transmitted from one person to another like an infection. Patients and animals suffering acute radiation syndrome can be treated without fear of the attendant catching the illness.

In Hiroshima and Nagasaki animals were exposed to radiation, and they showed the same symptoms of radiation sickness as did exposed humans. Prior to the atomic bombings of Japan and continuing to the present day, animals of all sizes and species have been used in radiation studies. The higher mammals from mice, rats, guinea pigs, rabbits, on up to dogs, sheep, horses, chimpanzees, and apes, show the same basic symptoms of radiation injury as human beings. In fact, laboratory test animals show primarily the same response to the various treatments for radiation sickness that humans show. This will be an important fact to remember when studying the chapters on preventive measures and treatment.

In discussing the acute radiation syndrome, it is necessary to refer to causative radiation dose levels. A *roentgen* is a measurement of radiation energy. A *rem* (roentgen-equivalent-man) is a measure of biological damage. A roentgen and a rem may be considered equal for the purposes of our discussion. That is, 100 roentgens equals 100 rems, and vice versa.

All living tissue, both plants and animals, is sensitive to ionizing radiation, that is, can be damaged by it. However, plants and animals vary widely in their sensitivity to radiation. For our purposes, there are two ways in which radio-sensitivity can be described. One method is known as the *Lethal Dose 50%* or LD_{50}. This means that taken as an average, the sensitivity of an animal species to radiation is expressed as that dose which is lethal (kills) fifty percent of that species. For example, the following table shows the full-body exposure dosage of radiation required to be lethal for different animals (50% of them).

The second, and less common method to express radiosensitivity, is the dose (in roentgens or rems) required to kill 100 percent of any number of animals in a species. It should be obvious that an exceedingly high dose rate, for example 50,000 r, will kill all the animals of all species in a very short time.

LD$_{50}$ OF VARIOUS ANIMALS EXPOSED TO HARD X-RAY DOSAGES (in roentgens)[2]	
ANIMAL	LD$_{50}$
Guinea pig	200-400 r
Swine	275 r
Dog	325 r
Goat	350 r
Monkey	500 r
Mouse	400-600 r
Rat	600-700 r
Hamster	700 r
Rabbit	800 r

For man, the lethal dose for 50% of the exposed individuals of a group (a city, etc.) is considered to be around 450 r, which is somewhere between a mouse and a goat. This 450 r dose is an average. However, the lethal dose will vary because of biological factors such as age, sex, and physical condition. The embryo and fetus are exceptionally radiosensitive. Children during the ages when rapid development (growth) is taking place in the bones and musculature structures are also highly radiosensitive. Women of child-bearing age are the next most radiosensitive group. Then come men up to approximately 40 years of age. Men and women 50 years of age and over are the least radiosensitive (or the most radioresistant). Since pregnant women (and their fetuses) and young children are particularly radiosensitive, they should be protected accordingly. An abnormally high incidence of deformed and defective children were born to Japanese women who were pregnant and exposed to the effects of the atomic bombings of Hiroshima and Nagasaki. Radiation exposure of pregnant women can result in babies with abnormally small heads, blindness, cataracts, retinal problems of the eyes, mental deficiencies, malformed skulls, cleft palates, spina bifida, deformed arms, club feet, genital deformities, and general and physical subnormalities, and central nervous system abnormalities.[3] There were, however, many Japanese women who were pregnant and exposed to varying types and intensities of radiation who delivered babies that were normal. The most damage to the developing fetus occurs from conception through the first five months of pregnancy.

Furthermore, there is a definite difference in the radiosensitivity of certain tissues in the human body. Rapidly proliferating tissues are the most sensitive to radiation. And within cells of tissues, it has been overwhelmingly demonstrated that the nucleus of a cell is the most radiosensitive.[4] In general, the blood and blood-forming tissues are considered extremely radiosensitive. The second most radiosensitive are the intestinal epithelium (mucous membranes) and germinal tissues, followed by the skin and connective tissues. Glands and bones are considered relatively resistant to radiation penetration from outside the body, but food and water contaminated by fallout particles containing radioisotopes of iodine or strontium may cause damage as internal radiation emitters to the thyroid gland and bones respectively.

Alexander Hollaender cautions in *Radiation Biology* that: "High energy radiations dissipate their energies in tissue by ionization and excitation. A relatively very small absolute amount of energy absorbed by tissue is required to produce widespread damage."[5] Radiation damage to one cell can have a wide-ranging effect on adjacent cells presumably due to release of toxic products or other factors. When living tissue is exposed to penetrating radiation, energy is absorbed fairly uniformly throughout the tissue mass, but the amount of radiation energy which will produce lethal effects in cells is exceedingly small. Even though it is true that there are great differences in radiosensitivity between different types of tissue, it is important to understand that there are no absolute or clearly defined categories of radiosensitivity.

Radiation-injured Tissue Regeneration

Again, quoting Hollaender, "A striking characteristic of living substance is the capacity to repair deleterious changes produced within it by external agents."[6] Tissues (in most instances) are able to regenerate after low level radiation exposure. But above certain radiation levels tissue does not retain the ability to regenerate. This depends upon the type of tissue. Since some types are more resistant to radiation than others, they are capable of regeneration after exposure to relatively high dose levels. "The regeneration of injured tissue may occur before the cessation of tissue destruction, or at some variable time thereafter. The degree to which a tissue regenerates depends upon the number of surviving stem (cells) or mother cells and upon the innate ability of the surviving cells to proliferate. It also depends upon the damage sustained by the vascular (blood) supply of the tissue .... What factors govern the time at which regeneration starts are not clear. Healing of the epithelium in an acute X-ray ulcer may be delayed for months and then occur suddenly and rapidly. Waves of destruction have been described in some tissues, and these phenomena also are beyond explanation at present."[7] So, while regeneration (healing) of radiation-injured tissue occurs, it is often generally at a slower rate than in non-radiation injuries of similar scale.

Cell Division and Tissue Damage

Body tissues grow and regenerate by cell division. In cell division (mitosis), the cells elongate, become smaller in the middle, then the nucleus seems to stretch within the cell, and in a short time there are two separate tissue cells. This process takes place continuously in living tissue, and in the human body it occurs millions of times every day. It takes place at a very rapid rate in certain tissue types.

The reactions displayed by body tissues exposed to radiation are due to damage to and death of cell components. There are two main types of radiation-caused cell damage. The first occurs when cells are in the resting stage (inter-mitotic stage) between cell division. This mode of cell death occurs soon after exposure to ionizing radiation, and is usually caused by relatively large doses of radiation. The second mode of cell death does not occur until fatally injured cells attempt to divide (mitosis). This type of cell death is produced by relatively small doses of radiation, and is presumed to be due to damage of the cell chromosomes. This second mode of cell death is an interesting phenomenon and is a peculiarity of ionizing radiation and a few mitotic poisons. It is also characteristic of burns caused by heat or electricity which result in immediate tissue destruction due to coagulation of cellular proteins. The total body responds to the damage to localized and superficially dead tissues. But in acute radiation syndrome there occurs both a localized and a general "systemic reaction to dead and dying cells which persist in ... tissues throughout the body for a matter of weeks. The prolonged period of tissue destruction is not unlike that in certain chemical poisonings or bacterial infections."[8]

Clinical Characteristics Shared by Radiation Damage And Chemical Poisons

1. Large single doses cause serious illness or death, depending on the size of the dose and individual susceptibility, except that very large doses are invariably fatal.

2. Small daily doses can be tolerated over a long period of time. The total amount received in this manner without causing illness may be many times greater than the size of the single lethal dose.

3. Combinations of large single doses and repeated small doses have intermediate effects.
4. The ability of the body to recover from a large single dose and to tolerate much larger amounts received as repeated small doses depends on such biological processes as repair of injury, elimination, etc.

Clinical Characteristics Shared By Radiation Damage and Thermal Burns

1. Massive and often rapid fluid loss occurs (dehydration) with consequent electrolyte imbalance.
2. There is an increase in susceptibility to infection.
3. Fever.
4. Probability of a response from the adrenal glands (secretion).
5. Amino acids are excreted (primarily in the urine) in large amounts.
6. In severe injury, intestinal obstruction is likely.
7. Shock, followed by possible circulatory collapse in severe injury.

Enzyme Inactivation

Body cells contain enzymes that are essential for cell metabolism. Enzymes are protein substances that act as catalysts in various chemical functions within the cell, especially reproduction and synthesis of blood nutrients. Ionizing radiation appears to affect body metabolism by acting upon enzyme systems and other cell components. Shock-like symptoms appearing in exposed individuals are due to toxic products released because of massive tissue destruction. The toxic products released by destroyed molecules cause additional tissue damage which may activate some enzymes and inactivate others. "It is known that the in-vivo (within the living body) activity of several enzymes is increased following irradiation. This may be said to be a general phenomenon of radiobiology."[9] It is also known that as radiation dose levels increase, urinary amino acids are excreted in progressively larger amounts. This is believed to be due to the inability of the damaged cells to use the amino acids and to progressive tissue destruction. "Inactivation of a single enzyme molecule by radiation may influence thousands of other molecules in the reaction system . . . . It has been calculated that at least 1,000 to 100,000 nucleic acid molecules are indirectly affected for every molecule directly ionized by radiation."[10]

Clinical Course of the Illness

The most effective method to prevent radiation sickness is to prevent exposure to radioactive sources. However, during and after a thermonuclear war complete prevention of exposure for most individuals will be impossible. Nevertheless, reduction of the extent of exposure is possible. It has been estimated that over 100,000 survivors at Hiroshima and Nagasaki received sub-lethal doses of ionizing radiation. The most outstanding characteristic of the response of the human body to exposure to ionizing radiation is the wide diversity of symptoms and signs that ensue. Following exposure, the function of every organ system of the body is disturbed to some extent. The severity of the disturbance depends on the type and quality of radiation, its intensity, and the parts of the body most heavily exposed. As noted above, the extent of damage to different organ systems is also dependent on their radiosensitivity. If an individual receives a dose of 400 rem of full-body exposure, the damage will be much greater than if he receives a dose of 50 rem for eight consecutive days (8 days x 50 rem = 400 rem). Richard E. King, in his paper, *Survival in Nuclear Warfare*, backs up this claim: "If doses are divided, probably much more radiation can be sustained with survival. We base this opinion on our experience in total body irradiation of patients with incurable cancer. We have given such very ill patients 100 r doses two or three times a week, for two or three weeks, or a total dose of 600 r to 800 r, with no symptoms of radiation sickness. From this, we would hazard the guess that one might probably receive divided doses of 100 r daily for two weeks, or a total of 1,400 r, and still survive."[11] Carrying this point a step further, a Civil Defense publication states that: "Acute effects would be modified considerably if the radiation dose were received over a long period of time. The body repairs some of the damage (perhaps up to 90 percent) if it is given time. For example, a whole body dose of 600 r or more in a short period of time, say four days or less, would be fatal to most humans. But the same total exposure would not cause death or any noticeable effects if it were acquired in small doses over a much longer period, say a year or more."[12]

It is extremely difficult to correlate exposure intensity, type and quality of radiation, variations in human sensitivity, and the effect of whether medical treatment is, or is not, received, such that one can state that a specific amount of exposure will result in an exact

RADIATION SICKNESS

amount of damage or cause specific symptoms. For our purposes, three median responses to specific amounts of radiation will be postulated to provide an explanation of the clinical course of radiation sickness. For simplicity these three categories will be used: (1) Mild Exposure, (2) Moderate Exposure, and (3) Severe Exposure.

MILD EXPOSURE: 0-100 rem

PROBABLE SEQUENCE OF EFFECTS ON HUMANS RECEIVING FULL-BODY EXPOSURE TO PENETRATING RADIATION

Exposure: 1-25 rem

No visible injury. There will be a significant (but not extreme) decrease in circulating white blood cells, beginning the first few hours after exposure. A significant decrease in the production of red cells (formed in the bone marrow) following an exposure of as little as 5 r may occur. It is possible that a slight increase in blood coagulation time may occur. The white cell count usually returns to normal value after the first 48 hours. Platelet and red blood cell counts take longer, but will return to normal after the first week.

Exposure: 25-50 rem

Definite blood changes. Rapid drop in circulating white blood cells—more pronounced than in exposures less than 25 rem. Drop in blood cell counts will begin almost immediately after exposure. Coagulation time of the blood will be prolonged over exposures of less than 25 rems. Blood cell counts usually return to normal values after ten days to 2 weeks. Some fatigue (more than usual) may be experienced by some individuals. No visible injuries will occur. No medical treatment is required.

Exposure: 50-100 rem

Blood cell changes, some injury, but no serious disability. Generally, following doses between 50 and 100 r, the decrease in circulating white cells will start in about 15 minutes and continues, reaching its low point at about 36 to 48 hours. At this dosage level the recovery of the white cell count varies between 2 and 100 days, depending on the actual dose level and individual sensitivity to radiation. Platelets and red blood cell counts usually return to normal sooner than white cells (but not within 2 days). White cell values are usually the last to return to pre-irradiation levels. At this dosage, recovery is accompanied by return to normal values of white cells. In a few very sensitive individuals some nausea and vomiting along with mild depression may occur, but this is not usual. Fatigue and weakness occur in many individuals. At the upper levels of this dose range, some individuals experience malaise (uneasiness or general discomfort or distress). After a few days, these symptoms (except for the blood problems) will disappear. During the third or fourth week, some nausea, vomiting, and mild diarrhea may return, but these symptoms are controllable with mild medication. Malaise is likely to return along with weakness and fatigue. These third and fourth week symptoms will usually disappear about the fifth or sixth week with no recurrence except for fatigue. Total recovery can be expected by the tenth week, except for white blood cell counts.

MODERATE TO MODERATELY SEVERE EXPOSURE:
100-450 rem

PROBABLE SEQUENCE OF EFFECTS ON HUMANS
RECEIVING FULL-BODY EXPOSURE
TO PENETRATING RADIATION

Exposure: 100-450 rem

Some symptoms of radiation sickness due to moderate exposure will appear in about 2 hours. Distress from symptoms will usually increase during the first day. These include:

—rapid onset of nausea and vomiting
—lack of appetite
—extreme thirst
—fatigue, drowsiness
—extreme weakness and possible prostration

Most of these symptoms will disappear toward the end of the first day (24 hour period) but may recur the next day.

Between the third and fourth day individuals feel much better, resume work and fulfill regular duties. During the next two or three weeks no symptoms will appear except for unusual fatigue at the end of the day, and a general feeling of uneasiness (malaise). Following exposure at the higher end of this dosage range, there are often immediate behavioral changes. Dejection and inactivity appear followed by loss of appetite. After a few hours these symptoms disappear. Temporary sterility may occur at the higher dosage levels.

Radiation Sickness 53

Immediately following exposure, moderate to serious blood changes begin to occur, depending on the dose. Depression of activity of blood-forming tissues is general and widespread. Beginning about 14 days after exposure, many will begin to lose their hair (epilation), but regrowth will begin around the fourth or fifth week.

Between 21 and 28 days after exposure, the nausea, vomiting, diarrhea, abdominal pain, and loss of appetite will recur with even greater intensity than experienced shortly after exposure. The body's immunity system will be severely damaged resulting in reduced resistance to infections. Hemorrhages may occur from the skin, mouth, nose, and intestinal tract. Cuts and skin breaks will heal slowly, or not at all. Extreme weakness and fatigue will return. Radiostomatitis (sore mouth) will be present in many individuals. The combination of sore mouth, nausea, vomiting, and diarrhea accompanied by loss of appetite and poor absorption of food, will result in rapid weight loss. Severe fluid loss leading to dehydration will affect many individuals due to hemorrhaging, vomiting, and diarrhea. Menstrual disorders will be common among women. Fever of varying intensity will be present in most individuals showing any symptoms (at lower dosage levels). Severely ill individuals may rapidly become emaciated, and have severe difficulty in swallowing. Many will observe rapid changes in taste sensation as the severity of symptoms increase. Those exposed to higher dosage levels will have dry mouths, saliva will be very thick and scant, and there may be difficulty in swallowing.

The duration of this phase of the illness depends on the dose received, and general physical condition of the individual. As a rule, the higher the dose level, the longer it will last. Approximately 70% will survive, but recuperation may take many months. This group will need close medical attention. In general, the longer the time interval between the illness which occurs shortly after exposure and the start of the toxic phase (second illness), the greater the chance for recovery. Also, the shorter this time interval, and the more severe the initial symptoms, the less likely the chances of survival. Many of the deaths which occur within this group will be due to overwhelming infections.

> **SEVERE EXPOSURE: 450 rems and above**
>
> **PROBABLE SEQUENCE OF EFFECTS ON HUMANS RECEIVING FULL-BODY EXPOSURE TO PENETRATING RADIATION**
>
> Exposure: 450 rems +
>
> The symptoms described in the table for moderate to moderately severe exposure will be the same for severe doses. However, of those individuals receiving between 450 and 500 rems exposure, more than 50% will die. The same symptoms appear in persons who have been severely exposed as in those in the moderately severe exposed group, except that the onset of symptoms is more rapid, the symptoms more severe, and most will die.
>
> If exposure to radiation has been extremely high (1,600 rems and above), the central nervous system is affected. Damage may appear in a few hours up to 4 days, depending on dosage, and is characterized by loss of consciousness, brain damage and swelling, followed by certain death. The higher the exposure dose the more rapid the onset of coma. In lower dose ranges (450 to 600 rems), if vomiting and diarrhea are immediate and persist over several days, the individual has probably absorbed a lethal dose. Irreversible bone marrow depression occurs in doses between 700 and 900 rems. In individuals receiving exposures in the range of 600 r or slightly above, fatalities approach nearly 100%. The sequence of effects is generally as follows:
>
> FIRST DAY: Malaise, nausea, and vomiting begin one to two hours after exposure, and continue for several days or until death. There is prostration, and a very rapid and severe fall in the number of white blood cells.
>
> NEXT FEW DAYS: Severe diarrhea accompanied by extremely high, rising step-like fever.
>
> UP TO TWO WEEKS (without therapy): Delirium resulting in death, which usually occurs within 2 weeks after exposure. Large purple spots appear on the skin, due to hemorrhages beneath the surface, and hair loss may appear shortly before death. Those few who survive will convalesce for about six months.

Physical and psychological reactions resulting from a nuclear war such as dietary changes, fear, stress, and anxiety may result in symptoms appearing in some people which appear to be characteristic of the acute radiation syndrome (diarrhea, vomiting). Many of the symptoms of the illness are also characteristic of infectious diseases. Therefore, simply because an individual manifests a few symptoms does not necessarily mean he has radiation sickness. Once exposure to ionizing radiation occurs, it is too late to alter the course of the illness. After exposure, treatment is primarily symptomatic. But, the prognosis is good for those victims who survive the initial shock and secondary stages. "Although there are a number of means by which tissues can be protected to some extent against the damaging action of penetrating radiations, there is no known method of saving doomed cells once injury has occurred. However, there is every reason to believe that the regenerating powers of most tissues are so great that considerable, if not almost complete recovery will occur if the individual survives the critical toxic stage."[13]

PROBABLE EARLY EFFECTS OF ACUTE RADIATION DOSES OVER WHOLE BODY

ACUTE DOSE	PROBABLE EFFECT
0-15 r	No obvious injury.
25-50 r	Possible blood changes, but no serious injury.
50-100 r	Blood-cell changes, some injury, no disability.
100-200 r	Injury, possible disability.
200-400 r	Injury and disability certain. Death possible.
450 r	Fatal to (approx.) 50%
600 r+	Fatal.

SUMMARY OF SYMPTOMS AND TIME OF OCCURRENCE IN RADIATION SICKNESS
(Based on full-body exposure)

Time after exposure	Lethal Dose (600 rem and above)	LD_{50} (400-450 rem) Kills 50% of those exposed	Moderate Dose (150-350 rem)	Mild Dose (25-150 rem)
First Week	Nausea & vomiting after 1-2 hours	Nausea, vomiting & headache, 2-3 hours.		
	No definite symptoms		No definite symptoms.	No definite symptoms.
Second Week	Nausea, vomiting, diarrhea, weakness, headache, sore mouth and throat. Fever, rapid emaciation, death, (mortality around 95%).	No definite symptoms: May be depressed, feel vague distress, uneasiness.		
Third Week		Begin loss of hair, loss of appetite, general feelings of discomfort. Fever, weakness, prostration. Severe inflammation of mouth, sore throat.	Loss of hair. Malaise, loss of appetite. Weakness and fatigue. Sore mouth & throat.	
Fourth Week		Paleness of skin. Diarrhea, nausea, vomiting, bleeding of gums.	Purple spots under skin. Paleness. Diarrhea, nausea, vomiting (mild & controllable). Headache, emaciation. Recovery very likely unless complicated by infection, poor health, or concurrent injuries.	
Fifth Week		Purple spots under skin. Rapid emaciation. Death, (mortality about 50%).		
Sixth Week		On the way to recovery.		Apprehension, general uneasiness. Fatigue. Possible weakness. Rapid recovery.

9. PRE-EXPOSURE PREVENTION AND TREATMENT

OVER the last thirty-five years scientists world-wide have engaged in research of chemical substances which may protect humans and animals against the effects of ionizing radiation. This research has been especially intensive in the United States, the Soviet Union, Japan, Germany, England, France, and Italy. To cite one example, in June, 1970, there was a large gathering of scientists in Moscow called the First All Union Conference on the Pharmacology of Anti-Radiation Preparations. Many such conferences have been held in both Eastern-bloc countries and in nations throughout the free world. If one spends a few hours in the library of a medical school, and knows how and where to look, the amount of research conducted on finding protective substances is staggering, and so is the complexity of that research.

Literally thousands of chemical substances and procedures have been tested, and some have been found to provide definite protection for the body against the effects of ionizing radiation. What is disturbing is that the general public has never been informed regarding these protective substances and procedures, even though several are quite simple, safe, and relatively inexpensive.

Though these substances will provide considerable protection, this does not imply that their use will prevent all radiation damage, but rather that the body will be significantly more resistant to radiation's harmful effects, or that the effects will be modified. If the dosage of radiation is sufficiently high, serious damage or death will result no matter what precautions are taken.

Optimum Health

A person must be free from disease and disabling defects to have an acceptable level of health, but the term *health* encompasses more than this. No two people are exactly alike though both may be in the range of normal health. Health is a relative matter, and within a group of healthy people a wide range of physical well-being will be found. Some will possess a constitution which provides almost limitless vitality, endurance, and resurgence, even though they do not practice the principles of a healthful personal lifestyle.

Many gradations are found between the extremes of a highly proficient constitution at one pole and an inadequate constitution at the other. A person's health is an overall condition of well-being to be evaluated collectively, not a matter of specialized development. A person with normal health may suffer a temporary illness, such as a cold or appendicitis, or he may suffer a disability, such as a bone fracture, and thus be temporarily outside the range of normal health. But recovery is generally quick to the normal healthy range. Every individual has the capacity to attain a personal optimum state of health. Attaining an optimum level of health often requires changes in lifestyle. Once reached, the optimum level of health should be maintained.

War and emergencies focus attention on physical fitness. An integral part of every person's ability to survive is his health at the time of crisis. This will be especially true during and after a thermonuclear war. An individual enjoying his state of optimum health (or very close to it) stands a far better chance of survival, if injured by thermal burns, blast, or radiation, than if in a state of poor health. During a nuclear war and its aftermath, conditions may require people to eat questionable foods or kinds different than they are accustomed to, drink contaminated water, work long hours, be exposed to extremes of temperature—all this in an environment in which contagious diseases are flourishing. It is easy to understand why a higher percentage of the surviving population will be those in excellent physical health when the crisis begins.

Fundamentals of Achieving Optimum Individual Health

1. Take an inventory of present health condition through periodic examinations (including physical, eye, and dental).
2. Care for body functions, including those factors affecting the skin, teeth, hearing, and vision.
3. Avoid products and habits known to be harmful to health.
4. Provide essential nutrition and vitality for the digestive system through scientifically established dietary practices.
5. Adapt physical activity to one's capacities and needs in order to obtain maximum physiological and mental benefits.
6. Adjust pattern of living to avoid extreme fatigue and exhaustion.
7. Provide adequate relaxation, rest, and sleep in the daily program for the specific requirements of the individual.
8. Adapt to physical hazards through a safety consciousness

adequate to anticipate hazardous conditions and practices.
9. Develop a positive, mature, mental, emotional, and social perspective.
10. Use available health resources appropriately and at the proper time.

Immunity and Immunization

Immunity to infectious disease refers to the ability to combat infectious organisms after they have invaded the body, and thus, to prevent them from producing disease. Because immunity is of varying degrees in each individual, the term is used in a relative sense—it does not necessarily mean absolute resistance to an infection. A person may have a high degree of resistance to a disease, but may nevertheless contract the infection if his body is invaded by a highly virulent strain of the infecting microorganisms. The degree of resistance depends upon the individual and the virulence of the invading disease organism.

The intact skin and mucous membranes of the body provide some protection against microorganisms of all kinds. Bactericidal and fungicidal acids naturally occurring in the secretions of the skin provide additional protection. The natural flow of saliva, tears, and excretion of urine mechanically remove bacteria from the body, and to some limited extent each is somewhat bactericidal. Nasal secretions, saliva and sputum from healthy persons contain many types of complicated sugars (polysaccharides) which are capable of slowing down the destruction of red blood cells by the influenza virus. They probably also provide considerable non-specific protection against the disease.

The thin film of mucous which covers the epithelium of the surface of the respiratory tract from the nostrils to the terminal bronchii of the lungs traps all sorts of microorganisms, after which they are swept away from the upper respiratory tract into the esophagus and from the lower respiratory tract up to the mouth where they are swallowed. In most cases the gastric secretions complete the destruction of the organisms. The cough reflex is another mechanism which dislodges large accumulations of mucous and foreign material from the bronchii.

White blood cells continue removing organisms from the blood even when a disease is progressing to a fatal termination. The antibody-producing mechanism of the body forms a part of our natural immunity. Many chemical and other constituents of our bodies form barriers to infection as a first line of defense.

Despite society's progress in sanitation and the treatment of diseases with a vast array of miracle drugs, vaccination still remains a strong weapon against many diseases. In some cases immunization constitutes the only effective method of controlling a disease. The importance of immunization will increase dramatically in the aftermath of a thermonuclear war because many of the public health and sanitation measures that have reduced the incidence of such diseases as typhoid fever and diphtheria will break down, and because living conditions will be conducive to communicability. Furthermore, conditions will reduce the resistance of the body to invasion by disease organisms, since the elements of healthful living (varied diet of fresh foods, clean water, exercise) will be unavailable. Fortunately, many of the diseases that can be expected to arise can be effectively immunized against through vaccination. The level of antibody production after vaccination (or, as in the childhood diseases mumps, measles, chicken pox, after exposure) generally remains high enough to provide protection against subsequent exposures to the disease organisms.

Clearly, it is necessary to secure all vaccinations normally administered *now*. The family physician or a local health department can provide the information, and discuss with you the value of additional vaccines against such rarely encountered (under normal circumstances) diseases as cholera and Rocky Mountain Spotted Fever. For most people, this will be an insurance call, since they have vigorously maintained their family's immunization schedule. No single act of preparation will be more rewarding.

Substances Beneficial in Pre-Exposure Time in Prevention of Damage by Ionizing Radiation

Once a thermonuclear war begins there will be large areas not immediately affected by radiation or radioactive fallout, but which will soon become contaminated. How soon will depend upon many variables over which we have little or no control. It is reasonable to believe that many individuals, families, and communities will have a brief span of time in which to perform a few simple, yet extremely important preventive measures designed to further fortify the body against the damaging effects of ionizing radiation.

Over the past forty years almost every conceivable drug has been tested for its applicability to the problems associated with irradiation. It has been demonstrated that "administration of physiological saline and antibiotics is beneficial . . . WHEN GIVEN AFTER EXPOSURE."

In addition, it has been proved that, "There are a number of substances which if given BEFORE EXPOSURE, will protect

PRE-EXPOSURE PREVENTION 61

against radiation effects. Of these, cysteine produces one of the most dramatic results, but glutathione, as well as vitamin C, pentose nucleotides and nicotinamide, also increases the life span and survival rate IF GIVEN PRIOR TO IRRADIATION."[1]

In discussing the use of preventive substances, included are their physiological reactions, dosages, and precautions. These pharmaceutical substances are available without prescription. Each has been carefully researched, has proven beneficial in reducing damage to the body from ionizing radiation, and has been tested on humans and animals exposed to varying dose levels of radiation. As will be shown, there is a fairly large body of scientific information which indicates that if properly used, a relatively high degree of protection in humans (and animals) can be attained against the destructive effects of radiation.

L-CYSTEINE

By definition, "Cysteine is a sulfur-containing amino acid which is a by-product of protein metabolism which has failed to become completely oxidized,"[2] in the body. L-Cysteine is a non-essential amino acid which is formed from the essential amino acid Methionine in the bodies of humans and animals. Methionine, along with L-Cysteine, provides the main source of sulfur in the diet. (L-Cysteine and Cysteine are virtually chemically equivalent; and L-Cysteine is the name by which the substance is known and can be purchased from pharmacies and supply centers.)

Cysteine was found early to provide protection against the effects of radiation. However, the mechanism by which this is done is not completely understood. At least three biochemical theories attempt to explain how cysteine achieves protection. The point made about irradiation in general in a paper entitled, *Cysteine Protection Against X Irradiation*, is important to keep in mind. "It is generally considered that many of the biological effects of radiation can be attributed to the activated water reactions (in body tissues) which result from irradiation."[3] Evidence strongly indicates that sulfhydryl compounds can diminish cellular destruction in X-ray exposure, and also in gamma ray exposure.[4] "Cysteine has been shown to decrease damage to the skin."[5] Cysteine significantly modifies the blood changes in irradiated rats, and may raise the threshold for radiation effects in general.[6] Even if given before irradiation it does not modify sensitivity to radiation exposure,[7] but cysteine with vitamin C improves the survival rate.[8] Cysteine injected into laboratory animals 15 to 30 minutes before irradiation has been found to provide optimal protection.[9]

L-Cysteine is an amino acid very similar in chemical structure to cystine. Pre-treatment with L-Cysteine reduces toxicity from gamma and X-irradiation, but cystine does not. L-Cysteine, after long storage, is slowly converted to L-Cystine, which is not considered effective as a pre- or post-treatment against irradiation.

Source: L-Cysteine is available commercially in tablet form without a prescription. It may be obtained at some pharmacies, health product stores, and chemical supply companies.

Shelf Life: Cysteine tablets should be stored in a cool, dry place. As noted above, L-Cysteine, after very long storage in hot and humid conditions, will slowly be converted to L-Cystine. Effective shelf life of Cysteine is estimated between 5 and 10 years. This depends upon storage conditions, especially temperature and exposure to light and moisture.

Precautions: A rare inborn error of metabolism which is hereditary results in a condition known as cystinuria, which affects the ability of kidney tubules to re-absorb cysteine, cystine, and other amino acids. Cystine is rather insoluble, and in cystinuria this amino acid crystallizes out in the urinary tract and kidneys, forming calculi (stones). Since cystinuria is rare and hereditary, cysteine may be used as a protective agent against radiation effects in all individuals, except those who know they suffer from the disease. Cysteine is normally excreted in the urine and feces except in people found to suffer from cystinuria.

Administration: Cysteine tablets should be taken orally with a liquid as soon as a nuclear attack appears imminent, or upon receiving news of an actual attack. It takes from 4 to 12 hours for the body to become flooded with free sulfhydryl groups, which provide the protection against radiation exposure.

Dosage: Adults: Two 500 mg. tablets daily for the first 7 days, then one 500 mg. tablet daily until radiation levels decrease to a point where it is safe to be outside for periods of 8 hours or longer. *Children:* Under 1 year to two years of age, 250 mg. daily (½ of a 500 mg. tablet) for the first 7 days. Crush one 500 mg. tablet, then use ½ of the powder in a liquid (water, juice, etc.) for administration. After 7 days, use 250 mg. (½ crushed tablet) every other day until radiation levels subside to the same safety level as for adults. In the event of very high radiation levels from fallout, etc., the dosage for both adults and children should be doubled and continued until radiation levels reach a safe level.

Cost: From 15¢ to 20¢ per tablet, depending upon the region in which you live and the outlet from which you purchase the tablets. Cysteine tablets are usually sold in 500 mg. strength, and in bottles

PRE-EXPOSURE PREVENTION 63

(dark colored) of from 30 to 60 tablets. The cost of a bottle of 30 tablets ranges from $4.50 to $6.30. A bottle of 60 tablets ranges from $9.00 to $12.60.

POTASSIUM IODIDE

It has long been recognized that iodine is an important nutrient to both man and animals, and an essential component in the formation of thyroid hormone in the body. In adults, the body contains from 20 to 50 milligrams of iodine, approximately 30 to 40 percent of which is stored in the thyroid gland. Iodine ingested from food, water, and pharmaceutical preparations is rapidly absorbed into the blood stream. Some is taken up by the thyroid gland where it is converted into thyroxine. Thyroxine is a hormone that exerts a stimulating effect on body metabolism. The output of thyroxine from the thyroid gland is controlled by the pituitary gland. If the amount of thyroxine in the body is insufficient, the basal metabolism is decreased, with the consequence that the individual's total activity slows down.

Iodine deficiency is the cause of goiter, which is an enlargement of the thyroid gland. Today, iodized salt sold in grocery stores provides sufficient iodine to prevent goiter. Prior to use of iodized salt, goiter was common in many parts of the United States. An excess of iodine in the diet is excreted in the urine and feces.

In thermonuclear war, the detonation of fission and fusion weapons will result in very large quantities of radioactive iodine 131 and 133 being distributed into the atmosphere, which will then return to earth as fallout. Both iodine 131 and 133 have an affinity for thyroid tissue. Iodine 131 has a half life of eight days and iodine 133, a half life of 22 hours. Therefore, these two radioactive substances will not constitute a long-term fallout hazard after thermonuclear explosions, but WILL BE THE MOST DANGEROUS INTERNAL RADIATION EMITTERS DURING THE FIRST 60 DAYS AFTER DETONATIONS STOP. "At the end of 80 days the iodine 133 present in the environment would only emit about 1/1000 as much radiation per hour as at the beginning of the 80 day period. Because of this rapid decay, a 100 day supply of potassium iodide for each family member should be sufficient if a nuclear war ... were to last no more than a week or two."[10]

Thyroid Block

The following concerning potassium iodide originated in an FDA document. When radioiodines "are inhaled or ingested, they rapidly accumulate in thyroid gland tissue and are metabolized into organic iodine compounds. These compounds could reside in the thyroid gland long enough to cause local radiation damage. Although a variety of chemical substances can block the accumulation of radioiodine in the thyroid gland, the Food and Drug Administration concludes that potassium iodide appears most suitable. Almost complete (greater than 90%) blocking of radioactive iodine uptake by the thyroid gland can be obtained by the oral administration of potassium iodide just before or at the time of exposure. A substantial benefit (i.e., a block of 50%) is attainable up to three or four hours after acute exposure. The FDA has concluded that most of the radioiodine not taken up by the thyroid gland is excreted in the urine within 48 hours."[11]

The paper goes on to point out that potassium iodide has long been used to treat other disorders, without unacceptable problems occurring. "FDA concluded that the risks from short-term use of relatively low doses of potassium iodide in a radiation emergency are outweighed by the risks when the thyroid radiation exposure above certain levels from radionuclides released into the environment is considered likely."[12]

The FDA further points out that iodized salt and kelp are not effective sources of thyroid-blocking iodine because of the small amount of iodine contained in them. If potassium iodide is not available to you during a radiation crisis, do not attempt to use tincture of iodine as a substitute. Free elemental iodine is not an effective blocking agent, and is poisonous if taken in quantities considerably greater than the small amounts consumed when tincture of iodine is used to disinfect contaminated drinking water.

Potassium iodide is available in a number of commercially marketed salt substitutes and in certain vitamin and food supplement products. It is a commonly prescribed medication for respiratory ailments, and is used as the main ingredient in non-prescription cough medicine. Potassium iodide would be particularly useful for protecting a population that was not able to be evacuated from a high risk nuclear attack area. It also would be useful to populations exposed to radioactive fallout.

A sobering note must be introduced, however: "The use of potassium iodide in a radiation emergency is not a panacea. It does not reduce the uptake by the body of other radioactive materials, nor provide protection against external radiation."[13] "After a nuclear attack, very few ... survivors would be able to obtain potassium iodide or to get advice about when to start taking it or stop taking it. In areas of heavy fallout, some survivors without potassium iodide would receive radiation doses large enough to destroy thyroid function before modern medical treatments would again become available. Even those injuries to the thyroid that result in its complete failure to function cause few deaths in normal times, but under post-attack conditions thyroid damage would be much more hazardous."[14] These comments simply point up the importance of being adequately stocked with this important substance *prior* to an actual emergency.

Availability: In many states potassium iodide tablets have been taken off the list of prescription medications and can be readily purchased in pharmacies. However, some states still require a prescription. Regardless of where you live, any adult can purchase potassium iodide in granular or crystalline form from chemical supply companies without prescription. USP grade or CP grade should be used.

Expiration: Potassium iodide tablets are expiration dated, as are nearly all medications. However, there does not seem to be any problem with taking expired potassium iodide tablets, or solution made with crystals or granules. Tablets are somewhat hygroscopic (attract moisture); chemically, it is a very stable compound and does not break down into other chemical forms.

Cost: One pound of chemical reagent grade potassium iodide granules costs around $35. However, it may be purchased in ¼ pound bottle size at much less cost. A 100 tablet bottle of 300 mg. strength costs around $10.

Preparation: Cresson Kearny has developed this method of preparation: "To prepare a saturated solution of potassium iodide, fill a bottle about 60% full of crystalline or granular potassium iodide. (A 2- fluid-ounce bottle, made of dark glass and having a non-metallic screwcap top, is a good size for a family. About 2 ounces of crystalline or granular potassium iodide is needed to fill a 2- fluid-ounce bottle about 60% full.) Next, pour safe, room-temperature water into the bottle until it is about 90% full. Then close the bottle tightly and shake it vigorously for at least 2 minutes. Some of the solid potassium iodide should remain permanently undissolved at the bottom of the bottle; that is proof that the solution is saturated. Experiments with a variety of ordinary household medicine droppers determined that 1 drop of a saturated solution of potassium iodide contains from 28 to 36 mg. of potassium iodide."[15]

Administration: Potassium iodide tablets usually are 300 mg., which is a little more than double the recommended adult dose. Potassium iodide has an extremely bitter taste, and it is best to take with a beverage that will mask the bitterness. Tablets should be swallowed along with fruit or vegetable juice or milk. Drops of saturated potassium iodide solution should be added to fruit juices, vegetable juices or milk, stirred, and drunk. Water may also be used if nothing else is available. The taste can also be covered by adding four drops of saturated solution to food, such as bread.

Dosage: To obtain a thyroid block against radioiodines: *Adults and children (over 1 year of age):* 130 mg. per day. If 300 mg. strength tablets are used, break one tablet in half. Each half will be approximately 150 mg., enough for two persons or 2 separate doses. The extra 20 mg. per dose will not be excessive. If saturated solution is used: Take 4 drops per day (in beverage or in a bread-formed pellet). *Infants (less than 1 year old):* Crush ¼ of a 300 mg. tablet between 2 spoons and mix it with formula or juice. If saturated solution is used, administer 2 drops per day, by adding it to formula or juice.

Storage: Store tablets or granular potassium iodide in a dark colored bottle with the cap tightly closed, in a cool, dry, dark area. The container cap should be non-metallic.

Shelf life: Many years, without ceasing to be an effective thyroid blocking agent. Tablets or the granular or crystalline form stores better than the saturated solution.

SIDE EFFECTS AND PRECAUTIONS:

Pregnancy: From the FDA report cited above comes this cautionary note: "Women who have been treated with ... 300 mg. potassium iodide three times per day or greater for any substantial time during pregnancy have been reported to give births to infants with enlarged thyroid glands, which sometimes may cause respiratory obstruction."[16] But the risk entailed in not blocking the thyroid far outweighs this potential problem. Another problem associated with iodide treatment comes about in children with cystic fibrosis. Again, the damage to the thyroid gland is a more serious danger, and such children should take potassium iodide during a radiation emergency.

The Neonate: The nursing newborn is potentially exposed to additional radioiodines because of the tendency for them to be taken up by the mammary glands. Therefore, substitutes may be called for. Immediately after birth, there is increased uptake of iodine by the thyroid, which also heightens susceptibility. It is also probably true that this results in part from heightened excretion of iodine by the gland. Again, the evidence suggests that administration of potassium iodide will benefit the child.

The Infant and Young Child: This group is at greater risk than mature adults because 1) Their thyroid glands will absorb the same

PRE-EXPOSURE PREVENTION 67

proportion of radioiodines as adults, resulting in a per weight greater dose and, 2) Their diet (especially milk) is more likely to be contaminated. As with all five of these special groups, preventive countermeasures (use of milk substitutes, uncontaminated foods, sheltering) along with the thyroid blocking agent potassium iodide, are recommended.

The Elderly: Though potassium iodide in thyroid-blocking doses in elderly patients has been reported to induce hyperthyroidism, there is some question whether the association is due to pre-existing thyroid disease. For the same reasons cited above, the use of potassium iodide in a radiation emergency is recommended for this group.

Non-Thyroid Side Effects:
1. Skin and Mucous Membrane Reactions: The ingestion of iodides has been associated with a variety of skin ailments from mild rashes and lesions to acneiform papules and pustules. "Persons with certain skin conditions, e.g., pustular psoriasis, dermatitis herpetiformis, or pemphigus may experience an exaggeration of the disease. Acne and eczema may also be exacerbated."[17]

2. Iodide "Mumps" and Miscellaneous Reactions: In patients treated for asthma with large quantities of potassium iodide (3.6 to 36 grams per day), 7 percent developed iodide mumps, i.e., painful swelling of the parotid and/or submaxillary glands. Other reactions reported include brassy or metallic taste, conjunctival irritation, headache, nausea, vomiting, and diarrhea.

The FDA report cited concludes that significant side effects from potassium iodide have been reported "following chronic administration of daily doses *far in excess* of those necessary for thyroid blocking in a radiation emergency."[18]

VITAMIN C (ASCORBIC ACID)

Vitamin C, or ascorbic acid, is the most familiar of all the vitamins. Its important role in nutrition has been known for many years. Normally the highest concentration of vitamin C is found in the adrenal glands and in the pituitary gland, and is distributed throughout the rest of the body. It is directly absorbed from the small intestines, with the absorption rate and diffusion to body tissues being rapid. Excretion of vitamin C in excess of body requirements is primarily through the urinary tract. If saturated with vitamin C, the human body is capable of storing sufficient quantities for several months. In many natural products, vitamin C oxidizes upon exposure to light and air. It has a sharp, pleasant acid taste. Decomposition begins when heated to 190-192° F. Considerable amounts of the vitamin are destroyed by the heat employed in the canning of foods. It is often used in foods as an antioxidant to prevent rancidity.

It is non-toxic in humans even when very large doses are administered over prolonged periods of time. Vitamin C is known to be involved in many metabolic functions: 1) It is required in the formation of the intercellular matrix (collagen, for example); 2) It regulates the production and stability of bone protein matrix, dentine, and cartilagenous tissue; 3) It is required for proper healing of fractures; 4) It is required to prevent abnormal permeability of the vascular system; 5) As a reducing agent, it is involved in oxidation-reduction reactions within body tissues; 6) Strong evidence indicates that it is required in the production of adrenal-cortical hormones; 7) It plays an important role in the metabolism of amino acids; 8) It possibly acts as a carrier in some intercellular hydrogen transfer system providing a regulating mechanism in oxidation-reduction potentials within cells; 9) It is antiscorbutic (prevents the dietary disease, scurvy).

Normal Minimum Daily Requirements of Vitamin C

Infants—10 mg.; children (1-5 years inclusive)—20 mg.; children (6 years and over)—20 mg.; adults—30 mg.

In vitamin C deficient patients, "wounds do not knit together firmly and have a tendency to break down .... During the normal process of healing and production of new connective tissue, ascorbic acid appears in very high concentrations in the wound area. In depleted subjects and persons with scurvy, whose wounds fail to heal, little or no ascorbic acid is found in the (wound) area."[19] The same problem occurs in capillary walls of patients deficient in vitamin C. The intercellular cement or ground substance (equivalent to bone matrix) is defective and the vessel walls are fragile and permeable. The increase in fragility and permeability are often observed as bleeding beneath the skin and bleeding of the gums.[20] Thus vitamin C is essential for the rapid manufacture of new collagen in the healing process of wounds. In patients suffering from chronic and acute infections, and in those with long-term inflammatory diseases, vitamin C is sidetracked from the plasma to storage sites in the tissues. Surgical procedures, thermal burns, and lacerations have the effect of diverting vitamin C from the plasma to the site of injury in high concentrations.[21] Vitamin C in large doses is prescribed for post-operative patients in order to enhance the healing of surgical incisions. Vitamin C is prescribed for burn patients. The administration of vitamin C in such cases is based on the observation that stress caused by both burn and wound trauma reduces the serum level of the vitamin because the body has diverted its supply to the site of injury as a mechanism of the healing process. Vitamin C is therefore seen as contributing to the healing process itself, and a necessary part of any emergency preparations. The table below demonstrates the common nature of thermal burns and ionizing radiation.

TISSUE DAMAGE AND SYMPTOMATOLOGY COMMON TO THERMAL BURN VICTIMS AND PATIENTS EXPOSED TO IONIZING RADIATION

1. Increased capillary permeability.
2. Decrease in body immunity.
3. Susceptibility to bacteremia.
4. Extensive tissue damage.
5. Loss of large amounts of fluids (containing the soluble constituents of blood plasma).
6. Shock (due to pain and/or fluid loss).
7. Vascular shock and collapse.
8. Hemolysis (liberation of hemoglobin from red blood cells).
9. Disturbed electrolyte balance.

Thus, while there is not as much direct evidence for the value of vitamin C when treating victims of the acute radiation syndrome as there is for other maladies, its shared properties with those other maladies makes very probable vitamin C's efficacy in treating radiation sickness.

In research involving twenty-seven patients with malignant neoplasms (cancers), each was given tablets orally, containing a combination of cysteine and ascorbic acid, three times a day immediately prior to radiation treatment. In none of these cases was there a patient who took a bad turn after irradiation therapy. All 27 cases were given pre-arranged doses within a fixed period of time.[22]

Because vitamin C is used in so many ways by the body, it is valuable in countering many of the effects of ionizing radiation. The experiment here described reveals this broad scope of benefits. "The protective action of ascorbic acid against ionizing radiation from an internal source of ^{32}P (radioactive phosphorus[32]) was studied, and it was determined that the administration of ascorbic acid to animals exposed to an internal source of radiation enhanced cellular metabolism. Moreover, ascorbic acid exerted a beneficial action on the haemopoietic (blood-forming) system, neutralized the loss of bone marrow tissue destroyed by ionizing radiation, and restored the percentage composition of the white cells to values similar to those observed in the blood of control animals."[23]

Foods available during and following a nuclear war may be deficient in ascorbic acid. To maintain some degree of protection against the effects of radiation, the ascorbic acid concentration of tissues should be maintained at reasonably high levels through supplementation. If radiation damage to tissues does occur, body requirements for the vitamin will remain high.

Dosage: It is recommended that adults and children 1 year of age and over receive a minimum of 100 mg. of vitamin C daily. Infants under 1 year of age should receive a minimum of 25 mg. per day. Vitamin C tablets or powder may be dissolved in formula or juice for administration to very young children.

Excess vitamin C is rapidly excreted by the body. Persons of all ages tolerate the vitamin in very large doses over long periods of time. Preventive treatment should begin immediately upon notification of a nuclear emergency. Administration of vitamin C should continue until radiation levels are reduced to safe levels. In the event that dietary deficiencies of vitamin C occur, use of vitamin C in small quantities (recommended daily allowances) should continue indefinitely or until vitamin C content of food supplies is again adequate.

Precautions: Mild gastro-intestinal disturbances may occur, usually at very high dose levels. "No toxic symptoms are observed in man following the administration of large doses of ascorbic acid. One to 6 grams have been given orally and intravenously" without side effects.

Shelf Life: Vitamin C is oxidized upon exposure to light and air. Breakdown of the vitamin is accelerated as temperature increases, and as exposure to light continues. The vitamin should be stored in a cool, dark place, in airtight containers.

Source: Vitamin C may be obtained from grocery stores, pharmacies, health food stores, supermarkets, chemical supply companies, etc. It is most easily obtainable. No prescription is required.

Cost: Vitamin C is quite inexpensive, but costs vary from brand to brand, and from region to region. Cost will also vary in accordance to the strength of the product. A bottle of 250 tablets of 100 mg. strength costs between $2.50 and $6.00.

10. TREATMENT OF RADIATION SICKNESSS

RECOMMENDED treatment for persons suffering from the acute radiation syndrome is described below. Immediate professional medical care, with close cooperation and supervision by physician and attending radiologist for planning overall patient management. Isolation of the patient to prevent infection, and to allow complete mental and physical rest, in conjunction with good nursing care and hygiene. Therapy is both supportive—directed at helping the individual to survive the toxic phase of the illness, and palliative—affording relief, but not cure. Present treatment consists of:

1. Blood and platelet transfusions.
2. Liberal use of antibiotics to prevent infection.
3. Administration of plasma and electrolytes as needed for maintenance of fluid balance and blood volume.
4. Sedatives and tranquilizers along with pain relievers are frequently necessary to control anxiety, pain, and ensure rest.
5. Administration of bone marrow suspensions or bone marrow transplants for combatting anemia and low white blood cell counts.
6. Antiemetics may be used to control nausea.
7. Bland diet with the addition of supplemental vitamins are necessary to offset weight loss due to loss of appetite, malabsorption, vomiting, diarrhea, and hemorrhages.[1]

This type of treatment is fine—for peacetime, when very few people are exposed to sufficient radiation to require treatment. But treatment of this kind will not be available either during or after a nuclear war (refer to Part I, Chapter 3). The hard fact is that only a few will receive proper medical care unless families prepare by learning to treat radiation sickness, by obtaining the few simple medications required, and by accepting the idea that reasonably good treatment CAN be provided at the family level, without the direction or aid of either a physician or nurse. It is amazing what individuals and families can do during times of crisis.

Practical Considerations

1. It must be emphasized that radiation injury should be watched for at least four weeks after exposure, especially when the size of dose is unknown.
2. Injured persons will have difficulty eating.
3. Transfusions and intravenous injections will be difficult to administer.
4. Low grade fever, occurring immediately after exposure, is restricted to fatally injured patients. Early fever is a response to tissue breakdown, while high fever is probably due, in part, to infection.
5. During the period of nausea and gastro-intestinal distress "force feeding is of little or no use ... because the damage to the intestinal lining interferes markedly with the processes of absorption. Absorption is also complicated by the depression of secretory activity of the gastric glands and reduction of hydrochloric acid, and excessive activity and spastic contraction of the intestines."
6. "Disturbances in water metabolism and distribution can be expected after irradiation as a result of vomiting, diarrhea, bleeding, modified fluid intake, etc., but these changes in water balance appear to be secondary and to not contribute significantly to the cause of death."[2]
7. Such treatment as is necessary should be provided to relieve physical discomfort of patients (headache, thirst, nausea, cleanliness, etc.).
8. Following the early symptoms of radiation sickness, hemorrhage, anemia, infection, and malnutrition appear. Severe hemorrhage is often the chief cause of death.
9. "Infection is one of the very most important parameters of radiation sickness, because of the loss of body resources to combat it and its effect on hemorrhages and general debility."[3]
10. Irradiated body surfaces are often quite sensitive to heat and sunlight. During treatment or convalescence, heat applications to irradiated body surfaces or sunbathing are to be avoided.
11. The patient knows best how he feels. Care and treatment should follow the axiom: Do the patient no harm.
12. Treatment and patient care should include psychological support.

Continuation of Preventive Medications

During the time of fallout and periods of dangerous residual radiation, and for at least four weeks afterward, it is important to

Treatment

continue the administration of preventive medications (potassium iodide, cysteine and vitamin C) as outlined in Part II, Chapter 9. It is possible that radiation damage may not be manifest up to four weeks following exposure, and during this period protection must be maintained. Maintaining optimum health during this time and thereafter will ensure preparedness against post-war hazards of disease and the rigors of a different type of living. At best, nearly everyone will receive some exposure from direct radiation or fallout.

Radiation sickness is treatable, and patients will respond to proper, even though austere, therapy unless fatal doses have been sustained. Just as in chronic illness and infectious disease, if the body sustains overwhelming damage, the disease will be fatal, even with the best of treatment. There is a higher fatality rate in untreated patients with radiation sickness as compared with patients who receive either total or partial medical care. Family care consisting of good personal hygiene, rest, and good nutrition complemented with supportive medications and treatment can increase survival to a remarkable degree. The regenerative power of most organ systems is so great that almost complete recovery will occur if the patient is able to survive the toxic phase of radiation sickness. In this respect there are several medications and treatment procedures that can be used as supportive and therapeutic measures for treating radiation sickness.

Infection and Resistance

Patients with radiation injury are highly vulnerable to micro-organisms. Even micro-organisms normally incapable of causing infection can establish an infection in a radiation-injured individual which can rapidly overwhelm the body. Radiation exposure suppresses the body's ability to ward off such infections. Wounds heal slowly and poorly, often with blood and lymph fluid oozing freely from superficial cuts and burns, thus enhancing the possibility of invasion by micro-organisms. Relatively high radiation doses cause destruction of the intestinal mucosa allowing invasion by micro-organisms in the intestines and the bloodstream. Depletion of body fluids due to diarrhea, vomiting, and hemorrhage further reduce the ability of the body to withstand infection. Gastrointestinal damage also reduces the patient's ability to absorb nutrition.

"All animals including man possess a certain degree of resistance to the organisms which cause disease. That is why only a small but varying proportion of the population which is infected (invaded), actually develops clinical symptoms of a disease."[4] Many of the micro-organisms normally found in different parts of the body, such as the nose, throat, intestinal tract, and on skin surfaces are not pathogenic under normal circumstances. A micro-organism is classified as a pathogen (disease producer) if it has the ability to incite disease in susceptible animals or humans. "Virulence is the term used to designate the disease inciting powers of a particular micro-organism and especially a specific strain.... There are organisms which have an intermediate position in the classification of virulence. The colon bacillus, for example, is a constant inhabitant of the normal bowel,"[5] but which, under certain circumstances, such as exist in patients suffering radiation sickness, can cause serious illness. When the body's defenses against infection (invasion by micro-organisms) are suppressed due to radiation damage, the normal flora of micro-organisms found on and in the body can invade the tissues. *Invasiveness* describes how rapid is the spread of organisms throughout the tissue of the host, while *virulence* is reserved for the property of causing death in an infected organism.

Antibiotic Therapy

Antibiotics are extremely effective in treating radiation sickness, chiefly because they assist in preventing infection from overwhelming the body during the toxic phase. The Japanese had very little with which to treat radiation sickness. The long years of war had depleted reserves of medicine and food for Japanese civilians, and the magnitude of destruction of the bombings was a complete surprise to the populations of Hiroshima and Nagasaki. "Once the American forces occupied Japan, medical supplies were provided by the U.S. Army, the International Red Cross, and the American Red Cross. Many of the medicines provided were not at that time widely used in Japan—such as penicillin, sulfadiazine, sulfaguanadine, and plasma; and these proved extremely effective for infections after the subacute phase."[6] Since then, antibiotics have been widely used in the treatment of radiation sickness, which has mainly occurred among patients receiving radiation therapy for various types of cancer.

Penicillin was the antibiotic of choice up to the late 1940s, primarily because of its availability. Since then, many other antibiotics have become available. Penicillin has been found to have a number of undesirable, sometimes even fatal, side effects. Many persons have developed extreme sensitivity to it, and a number of micro-organisms have developed resistance to it. Still, it can be used very effectively in the treatment of radiation sickness, and if available should be used if other more effective antibiotics are not available. As a precaution, patients must be asked if they are allergic to penicillin prior to its administration. Extremely severe anaphylaxis can occur due to penicillin allergies, which can be rapidly fatal. This condition is not at all uncommon in our population.

TETRACYCLINES—TERRAMYCIN AND AUREOMYCIN

The antibiotics of choice in treating radiation sickness are aureomycin and terramycin. Both are wide-spectrum antibiotics, that is, they have the ability to inhibit the growth of a wide range of micro-organisms. They both are particularly effective in controlling diarrhea and infections associated with the illness. They are also effective against the organisms which cause rickettsial infections, and a few of the viruses (lymphogranuloma and psittacosis).

Aureomycin is known chemically as chlortetracycline. *Terramycin* is known chemically as oxytetracycline. Both are highly soluble in water, rapidly absorbed by the body, and are quite stable compounds when compared with other antibiotics. These qualities make terramycin and aureomycin excellent antibiotics in treating radiation sickness in humans and animals.

Availability: All tetracycline drugs, including terramycin and aureomycin, are prescription antibiotics, and therefore are obtainable through pharmacies with a prescription from a licensed medical practitioner. Because the manufacture and distribution of these antibiotics will be disrupted or ended during and following a thermonuclear war, quantities should be obtained well in advance of such an event. They will urgently be needed.

Veterinary tetracyclines (terramycin and aureomycin) and other antibiotics may be purchased without prescription for treating livestock and poultry infections. They may be purchased at veterinary supply outlets, farmers cooperatives, and drugstores. Veterinary antibiotics may be used to treat livestock suffering the effects of radiation sickness. As noted above, the symptoms of radiation sickness in mammals are the same as those of people. There is little or no difference in the manufacture, quality control, and safety of veterinary antibiotics and those packaged for human consumption, except for directions for use. If the only antibiotics available were those intended for veterinary use, and their use meant the difference between life or death, few people would hesitate to use them. In the treatment of radiation sickness, probably the single most important part of patient management is antibiotic therapy.

Precautions: Terramycin and aureomycin are not to be used by individuals with a history of hypersensitivity to any of the tetracyclines. "Use of tetracycline class drugs during the last half of pregnancy, infancy, and childhood to the age of 8 years may cause permanent discoloration of the teeth (yellow-gray-brown). This adverse reaction is more common during long-term use of the drugs but has been observed following repeated short-term courses. Enamel hypoplasia (defective or incomplete development) has also been reported. Tetracycline drugs, therefore, should not be used in this age group unless other drugs are not likely to be effective."[7] Tetracycline should not be used in individuals with kidney impairment. Other side-effects may include skin sensitivity to direct sunlight or ultraviolet light and toxic reactions in the developing fetus. As with other antibiotics, the use of aureomycin or terramycin may result in proliferation of non-susceptible micro-organisms. Gastrointestinal side-effects may include: loss of appetite, nausea, vomiting, diarrhea, inflammation of the tongue, and difficulty in swallowing. Other side-effects are quite rare. There are far fewer side-effects from the use of the tetracyclines than from the use of penicillin, yet penicillin is more commonly used.

It is acknowledged that there are certain problems with the tetracyclines as with all antibiotics. First, some micro-organisms have become resistant to them over the period they have been used. Second, they are not effective against two of the most likely infections to be found following a thermonuclear war and its attendant breakdown, staphylococci and streptococci. There are two other classes of antibiotics that, if possible, would be useful to have on hand. Both penicillins and sulfa drugs are effective against those two prevalent micro-organisms. Sulfa drugs also are extremely effective in treating urinary tract infections. The advantage of tetracyclines recommended above over these two other classes of antibiotics are: (1) tetracyclines are *available* and, (2) treat a wide spectrum of potential infections. Penicillins have the shortcoming of not lasting long (especially unrefrigerated) but obviously, if one has some on hand, or could obtain some (by prescription), then it would be wise to do so. Fresh penicillins would remain effective through the period of greatest danger.

Dosage: Consult your family physician and follow label and packaging directions.

Aureomycin and terramycin have an extremely bitter taste—natural or artificial fruit juices or flavored beverages mask the bitter flavor. The same precautions apply in the use of veterinary tetracyclines as for those prepared under physician's prescriptions.

Tetracyclines should be administered for a minimum of 48 hours after all symptoms have subsided.

Antacids or other medications containing calcium, magnesium, or aluminum should not be taken at the same time with tetracycline therapy, since these substances inhibit tetracycline absorption. Dairy products and certain other foods also interfere with the absorption of tetracyclines.

Orally administered tetracycline should be given one to two hours after meals, and should not be given with milk or any other beverages containing dairy products. Oral dosages of tetracycline to infants should not be given with milk or formulas containing dairy products.

Splitting the daily amount of tetracycline into four equal doses provides a more constant level of antibiotic in the tissues at any given time. This is more desirable when fighting infection than one very large single daily dose. Smaller doses are usually tolerated better because the predominant side-effect of tetracycline therapy is nausea.

Shelf Life: Shelf life of the tetracyclines is approximately five years without serious loss of potency. Terramycin is somewhat more stable to the effects of heat than aureomycin, but both should be stored below 77° F. Cooler storage temperatures prolong shelf life. Evidence indicates that extended shelf life increases the likelihood of side-effects of tetracyclines.

ANTI-EMETICS

Anti-emetics are drugs which prevent or ease the presence of nausea. Some are available only with a prescription, others (dramamine, etc.) are available without prescription. "Symptoms from radiation sickness...can be controlled in part by an anti-emetic (e.g., prochlorperazine 5-10 mg. orally or intramuscularly three times a day) and may be prevented by its prior administration."[8] Smaller, more frequent meals may help to control nausea, if the patient can tolerate foods. During nausea or gastro-intestinal distress, force feeding of patients is not recommended. Nausea and vomiting are two of the first visible symptoms of radiation injury. However, during crisis situations, the actual cause may as easily be psychological: fear, apprehension, the sight and smell of seriously injured people, or simple upset stomach due to excitement. Nausea and vomiting must be accompanied by other symptoms before a diagnosis of radiation sickness can be fully justified. Remember that, as Herman Kahn vividly described it, "if you get a fatal dose of radiation the sequence of events is something like this: first you become nauseated, then sick; you seem to recover; then in two or three weeks you really get sick and die. Now just imagine yourself in a postwar (or post attack) situation. Everybody will have been subjected to extremes of anxiety, unfamiliar environment, strange foods, minimum toilet facilities, inadequate shelters, and the like. Under these conditions some high percentage of the population is going to become nauseated, and nausea is very catching. If one man vomits, everybody vomits. Almost everyone is likely to think he has received too much radiation."[9] No matter what the cause of a person's nausea, he should go to bed. Give him warm liquids only when they can be tolerated. Soft, bland foods should be eaten after liquids have been retained for 24 hours.

If intractable, continuous nausea and vomiting persists for two days, it may be assumed that radiation injury is the cause. Even so, the treatment for nausea and vomiting should be continued along with other supportive medications and procedures (antibiotics, salines, etc.). Weakness and severe headache are frequently associated with the vomiting, nausea, and diarrhea of radiation injury.

Precautions: Drowsiness is often associated with anti-emetics, and precautions should be taken against operating motor vehicles or other machinery which may prove to be dangerous.

Dosage: Follow package directions for each particular anti-emetic.

Availability: Can be purchased in drug stores and grocery stores. It is wise to request information from your pharmacist when deciding upon an anti-emetic.

Shelf Life: Refer to product label.

VITAMIN SUPPLEMENTS

Nausea, vomiting, loss of appetite, and hemorrhage, plus malabsorption of nutrients due to damage of the intestinal mucosa, all play a prominent role in malnutrition problems associated with radiation sickness. A prominent feature of this illness is weight loss. Therefore, the amount of vitamins and amino acids available to the patient, plus his ability to absorb them, will play a vital role in surviving the toxic phase. Supplemental amino acids and vitamins administered to the patient, especially prior to the development of the toxic phase of the illness, could mean the difference between survival and death.

Vitamin supplements, in addition to vitamin C, can help make up for deficiencies in the vitamin content of stored food. Vitamin supplements are also prescribed as part of the treatment for second and third degree burns and, as has been pointed out, thermal burn injuries and radiation sickness have many commonalities. Vitamins must be obtained from the diet and cannot be synthesized by

TREATMENT 79

the body. Therefore, since their functions are catalytic in cell metabolism, and they can only be gained either through diet or diet supplementation, it is wise to maintain a stockpile of them.

A number of symptoms of the acute radiation syndrome parallel symptoms of avitaminosis (disease due to deficiency of vitamins in the diet). Because of the problem of weight loss, malnutrition, and malabsorption of nutrients, bleeding, similarities to burn injury, etc., continuous vitamin therapy is indicated during the course of radiation sickness. It is recommended that a supply of multiple vitamins sufficient for one year for every family member be obtained and stored. The multiple vitamin supplements which are chosen should be able to be conveniently administered to children and adults on a daily basis. Because of the multitude of vitamin products available on the market, it would be wise to consult your pharmacist or physician regarding the best choice for your specific family needs.

Availability: Grocery stores, drugstores, health food stores, and other retail and wholesale business outlets. May be purchased without prescription.

Precautions: Refer to labeling and/or insert accompanying the multiple vitamin product you purchase.

Dosage: Consult with pharmacist, physician, or refer to label directions.

Shelf Life: Refer to label. Most vitamin supplements are dated regarding expiration. However, even if the expiration date has long since passed, considerable potency remains. If stored with lids tightly closed, in a cool, dark place, potency will remain relatively high.

Cost: The cost of multiple vitamin supplements varies greatly according to brand, region in which purchased, and potency.

AMINO ACIDS

Proteins play an important role in all metabolic functions, and are made up of amino acids. Of the 22 known naturally occurring amino acids, nine are essential for growth in children; 8 are needed to maintain nitrogen equilibrium in adults. Nausea and vomiting, plus malabsorption due to damage of the intestinal mucosa play a prominent role in the nutritional problems of irradiated patients. The amount and/or absorption of amino acids may be diminished as well. This points up the need for amino acid supplementation.

In a study of three patients who were accidentally exposed to large amounts of ionizing radiation in 1945 during development of the atomic bomb, each had their urinary output of amino acids closely monitored and compared with that of a (control) group of normal, healthy persons that had received no radiation exposure. "In each case, abnormal amounts of amino acids were observed in the urine sample collected during the first day following exposure. The amount of amino acids increased, reaching a peak on the sixth day after exposure. After the twelfth day following exposure (of 2 cases, one having died), the amino acid content of ... urine decreased rapidly, although traces of abnormally occurring amino acids were still evident during the fourth week after exposure. Besides the sustained amino-aciduria in these two patients, an inconsistent increase of individual amino acids was noted in the other survivors of a second accident shortly after exposure."[10] "Amino-aciduria (increased presence of amino acids in the urine), presumably due to a decreased utilization of amino acids and possibly to increased protein breakdown, was pronounced."[11]

Amino acid preparations are recommended to enhance protein synthesis in the body. Although it is important to maintain adequate concentrations of amino acids in the body for optimum health, it is of much greater importance in cases of radiation exposure. Rapid excretion via the urinary tract requires constant replenishing.

Amino acid tablets and powder are readily available without prescription, but it is important to obtain ones which contain sufficient quantities of all essential amino acids for daily use. Consult your physician or pharmacist.

Precautions: None.

Dosage and Administration: Adults and Children (Ages 5 years and above): Follow label directions. *Infants and Young Children* (Up to 5 years of age): One-fourth of the adult dose. Tablets may be crushed and added to milk, formula, or fruit and vegetable juices for infants, or powdered amino acids may be added. Children above one year of age can either swallow ¼ of an adult dose or use amino acid powder, according to individual preference.

Shelf Life: Refer to label. Many are not dated. Tightly closed caps, and storage in a cool, dry place prolongs potency. In all probability the shelf life is several years.

Availability: Available without prescription from drugstores, health stores, and some grocery stores.

Cost: Cost of amino acids varies according to brand, quantity, and manufacturer. Approximate cost per 100 tablets is around $8.50.

MINIMUM DAILY REQUIREMENTS OF THE ESSENTIAL AMINO ACIDS FOR ADULT MEN AND WOMEN.[12]

Essential Amino Acid	Daily requirements in grams per day.	
	Women	Men
Arginine	Non-essential for adults, essential for infants and young children.	
Histidine	Non-essential for adults, essential for infants and young children.	
Isoleucine	0.45	0.70
Leucine	0.62	1.10
Lysine	0.50	0.80
Methionine	0.55	1.10
Phenylalanine (& tyrosine)	1.12	1.10
Threonine	0.31	0.50
Tryptophan	0.16	0.25
Valine	0.65	0.80

Pain Relievers

The control of pain, and its alleviation, is among the most important aspects of medical therapy. Many of the types of injuries associated with exposure to radiation are accompanied by severe pain. Burns are of course very painful; and nausea is greatly discomforting. In many cases, the healing process is enhanced through the control of pain, because the patient's attitude is improved and other therapies become easier to administer. Clearly, it is reasonable to have in shelter as much medicine as possible, including pain killers, if available. Aspirin, and its sister drug, acetaminophen (commonly sold as Tylenol), are effective against a wide range of ailments. Prescription drugs that are more effective are also difficult to obtain. Consult with a physician to see whether there are alternative substances better suited to your needs, and to determine safe dosage levels.

Fluid and Electrolyte Balance

Water is the most abundant constituent of the body and plays a major role in cellular metabolism. A number of chemical compounds when dissolved in water dissociate and form ions. They provide for a continual interchange between the blood and the tissue fluid by the process of diffusion and are known as electrolytes. The amount of chemical substances forming electrolytes is very important for maintaining normal body functions, and must be kept in careful balance.

Increased body requirements for fluids and electrolytes are associated with abnormal losses due to vomiting, diarrhea, and bleeding in radiation sickness, which may result in serious dehydration. If the depletion is replaced by water alone there is consequently a further decrease in the electrolyte concentration of body fluids which may lead to shock.

Early therapy for all patients suffering thermal burns or radiation injury should consist of electrolyte fluids by mouth. In burn patients the effects on tissue surfaces are seen almost immediately, whereas in radiation injury the damage is not visible, and is characterized by later development of symptoms, and sometimes by successive waves of ulceration and repair.

There are several excellent preparations available without prescription for maintaining body fluids and minerals. They provide not only water, but a balanced combination of essential electrolytes including sodium, potassium, calcium, magnesium, chloride, citrate, and dextrose. Some are available in liquid form and some in powder form. They have long storage life and are inexpensive.

In an emergency a mild saline solution can be prepared by adding one level teaspoon of table salt to a quart of water.

A more effective solution can be prepared by adding one teaspoon of table salt and one-half teaspoon of bicarbonate of soda (sodium bicarbonate) to one quart of water.

Patients should be encouraged to take liquid as soon as possible, and should be started even though attacks of vomiting continue to occur. When vomiting has ceased the patient should be taking at least 6 to 8 cups of liquid a day. Either of the above solutions can be alternated with fruit juices and bouillon. If retained and absorbed, they aid in the treatment of shock; if not absorbed, but instead vomited, the patient's electrolytes are not further depleted. Thirst of the individual will, in most instances, dictate the need for the amount of fluids to be given. Patients should be urged to drink as much as they can tolerate, but fluids should not be forced.

Nutrition

In Japan, malnutrition as a result of food shortages during the war played an important role in the clinical history of the survivors of the atomic bombs. Under the very poor nutritional conditions following the war, physical recuperation of exposed victims was extremely inhibited. In addition to the problem of food shortage, frequently there is serious difficulty in maintaining sufficient caloric intake in radiation victims. As we've already seen, gastrointestinal distress diminishes food intake, and damaged intestines do not absorb nutrition as well.

Loss of appetite, nausea, vomiting, and sore mouth (radiostomatitis) combine to cause the patient to refuse food. There is often a decrease of saliva flow, resulting in a dry mouth. What saliva is secreted is sparse and very thick. Frequently the mouths and throats of patients are sore. Therefore, what food is taken may have to be washed down with large amounts of fluids. Complicating the ingestion of foods and liquids is the common problem of difficulty in swallowing. There is also the possibility that patients may develop a change in, or loss of, their sense of taste, which improves as they convalesce. Lack of saliva, or thick sticky saliva, can be most bothersome to some individuals. The administration of artificial saliva, which may be purchased at drugstores (without prescription) is very soothing. The administration of a few drops of artificial tears into the mouth also is beneficial, if artificial saliva is not available. Both artificial saliva and artificial tears contain methyl cellulose, which is prepared from wood pulp, and is used as a thickening agent in cosmetics, and in water soluble adhesives. Exposure to moderate and severe dosages of radiation results in progressive weight loss, partly due to extensive loss of body fluids, but mainly because of loss of appetite and reduced absorption of nutrients. "It is well known that the clinical response is influenced by the nutritional status of ... the person undergoing radiation therapy."[13] Maintenance of adequate nutrition is of prime importance in radiation sickness therapy.

PATIENT DIETS

The diet of choice for patients with radiation sickness "should be moderately high in calories and high in protein and vitamins, especially vitamin C," which is the same as prescribed for burn victims. Patients undergoing radiation therapy are frequently placed on "a high protein, high carbohydrate diet ... to maintain stable weight. A bland diet with frequent small feedings may be recommended to a patient experiencing nausea. A low residual diet is suggested to patients with diarrhea. Commercially available diet supplements are good for patients having difficulty swallowing or maintaining a stable weight."[14]

Caloric intake may be increased considerably by the use of supplementary sugar or honey dissolved in water or fruit juices.

Soft Diet

The soft diet is between a full liquid diet and a light general diet. It is easily digested, and causes little irritation when the patient has a sore mouth and throat. Only soft foods are included in the diet. Fried foods and those which contain significant amounts of indigestible crude fiber, or which are strongly flavored, are not included. Also excluded are raw vegetables, fruits, nuts, and whole grains.

Included	Excluded
—milk and milk products —soft drink, fruit juices —coffee, tea —precooked infant cereals & cereals not containing bran —fine, enriched white bread —ground or tender meat, fish and poultry (stewed, baked or roasted) —potatoes (without skins and boiled) —noodles —butter, margarine, oils —soft cooked or strained carrots, beets, peas, squash, spinach, tomatoes, tomato juice, sweet potatoes —soft cooked or canned pears, apricots, apples, peaches —soft bananas —soups (containing no "excluded" foods) —bland custards, puddings, ice cream —honey, jelly, spices, vinegar —gravies and creamed foods	—all other beverages —whole grain cereals —whole grain baked or fried foods —smoked, fried or salted meat, fish or poultry, lunch meats, sausage, bacon —aged or cured cheese —potato chips, potato skins or fried potatoes —all raw vegetables —all raw fruits (except soft bananas) —pies, doughnuts, pastries, spiced baked goods, or those containing nuts and fruit —berries —hard candy, or candy containing nuts or fruit —popcorn, nuts, marmalade, pickles

Bland Diet

A bland diet is one which is non-irritating to the gastrointestinal tract. It has a smooth consistency and texture. It does not contain foods with coarse fibers, seeds, skins, or cellulose. It is usually cooked, or pureed. The hard to digest parts of meat are reduced by trimming, or the meat is cooked with moist heat until tender. Bland diets contain no fried foods or strong condiments or spices. Such strongly flavored vegetables as onions, radishes, cabbage, brussel sprouts and cauliflower are not part of a bland diet. Coffee, tea, and alcoholic beverages are also not appropriate.

Foods included in a bland diet are:

—soft cooked peaches, apples, apricots, and pears	—poultry
	—fish
	—puddings
—soft cooked or strained carrots, beets, peas, spinach, sweet potatoes, squash, and tomatoes	—ice cream
	—soup
	—milk
	—eggs
—finely cooked and infant cereals	—meat
	—noodles
—citrus and fruit juices	—butter and margarine
—creamed vegetables	—bread
—unspiced cake	—cottage cheese

Low Residue Diet

Low residue diets are prescribed in conjunction with treatment of diarrhea and gastrointestinal disturbances. A low residue diet leaves a relatively small amount of residue in the lower intestinal tract after digestion and absorption of nutrients. Better absorption occurs if foods of this diet are administered in small amounts several times each day.

Included	Excluded
—poultry (lean)	—all fruits and vegetables
—fish (lean)	—fatty foods
—hard boiled eggs	—potatoes
—rice	—soft boiled eggs
—gelatin and jello	—butter, lard
—strained fruit juices	—swiss cheese
—sugar and honey	—milk and cream
—strained vegetable juices (thin)	
—decaffeinated coffee	
—cottage cheese	
—noodles	

DIETARY MANAGEMENT FOR PATIENTS SUFFERING RADIATION INJURY

The symptoms which are manifested by individuals suffering radiation injury often present a host of problems, which taken together, make nutritional management quite difficult:

1. Nausea, vomiting, and diarrhea deplete the body of fluids containing essential electrolytes, proteins, and amino acids.
2. Damage to the intestinal mucosa causes nutrients to be absorbed slowly and poorly.
3. Hydrochloric acid, normally secreted by the stomach to assist in the digestive process, may be inhibited.
4. The mouth, gums, and throat are often very sore and tender, which makes chewing and swallowing difficult.
5. Usually only small amounts of foods can be tolerated at a single feeding.
6. Dairy products, or foods made from dairy products, inhibit the absorption of the tetracycline antibiotics, which are an essential first line of defense in preventing infection.

The patient knows the most about how he feels about food intake at any given time. Patients should be encouraged to take liquids such as broths, thin soups, and especially mild salt solutions as often as possible to assist electrolyte balance. Prevention of dehydration and circulatory collapse takes precedence over food intake.

Force feeding of patients should not be done. It will do no good, and actually may cause additional damage. Multiple vitamin supplements, amino acid supplements, antibiotics, potassium iodide, and fluid intake are far more important than food intake during the acute toxic phase. When the nausea, vomiting, and diarrhea begin to subside, nutritional management should be secondary only to antibiotic and fluid therapy.

Diets for radiation injured victims must meet the following criteria:

1. High in calories.
2. High in protein content.
3. Soft enough not to be overly-irritating to the sore mouth and throat.
4. Contain enough liquid for ease in swallowing.
5. Easily digested and rapidly absorbed.
6. Bland (non-irritating).
7. High in salt content.

TREATMENT 87

8. Free of foods which interfere with antibiotic therapy (or other essential therapy).
9. High in vitamin and amino acid content.

Above all else, let the patient determine when to eat, what to eat (within limits), and how much to eat. If tetracyclines are being administered, dairy products are to be excluded from the patient's diet.

RECOMMENDED FOODS FOR RADIATION INJURED PATIENTS

—jelly —honey —sugar —butter —margarine —salt —noodles —soda crackers —rice	—fruit juices with extra sweetening —vegetables, pureed, or cooked tender, served with butter or margarine and salted —eggs (boiled), served with butter and salted —meats (red meats—especially liver; fish, poultry), ground or pureed, served with butter or margarine and salted —jello, or gelatin, with added sugar or honey. —bouillon soups, beef, chicken, etc. —creamed soups and vegetables (tender or pureed), served with butter or margarine, and salted —white bread with butter or margarine —potatoes, boiled (or instant flakes or nuggets), served with butter or margarine, and salted —puddings (made with very little or no dairy products) —plain cake or cookies —white cereals or baby cereals (no bran or roughage) Serve with butter or margarine, salt, and with nondairy creamer, or other imitation milk product

Nursing Care

The nursing care of patients with radiation sickness is similar to that required for patients suffering from burn injuries. Experience has shown that there are also likely to be many who suffer from wounds and burns who will require care. Basically, nursing care should be that found in normal situations. Though thermonuclear war is certainly not a normal situation, the basic fundamentals of good nursing care would still apply. Modern warfare may result in conditions which will expose us to dangers beside radiation, burns, and wounds. For example, family members may be exposed to weather extremes (heat or cold). Other conditions may cause fatigue, expose us to infection, or deprive us of food and water.

At the family level, the mother will be the one most likely to provide nursing care, but other family members should also be able to pitch in. What if the mother is the one that becomes sick? A Red Cross publication makes this point well. "Whenever there is a sick person in the home, it is beneficial for the patient if other members of the family cooperate in giving help. This is particularly true of young people, who, if they have some understanding of the needs of the sick, can contribute toward the patient's well-being and at the same time gain experience and satisfaction through carrying some of the responsibility."[15] A warm and friendly atmosphere with support from other family members is a positive aspect of home nursing care usually not found in hospitals, and which can significantly contribute to the patient's recovery. Often "the development of a healthy mental attitude may be as important to the recovery of the patient as the required physical care."[16] Family members should do their utmost to cooperate in developing a state of mental and physical tranquility by keeping conversations and activities restful and relaxing. Extreme anxiety and fear will only complicate the crisis, and cause undue strain on all family members. Psychological reassurance is important under stressful circumstances, but especially so for those who are sick. Good nursing care and hygiene are among the most important therapeutic measures available when treating radiation sickness. "Nursing care is the most important single phase of the treatment of radiation sickness."[17] Home therapy and nursing management should be planned for in advance. Thus, antiseptics, antibiotics, sedatives, analgesics, anti-diarrheal and anti-emetic drugs, table salt, bicarbonate of soda, nose drops, cough syrup, etc. should be available.

Patients suffering from radiation sickness and other injuries and illnesses will be more likely to survive if they are relatively free from stress, receive adequate nutrition, rest, and have clean surroundings. Radiation injured patient management requires that "Skin, oral, and nasal hygiene must be maintained at high standards. Anal hygiene is extremely important."[18] Many patients, even without such advantages and little or no medication, will survive. This was proven after the atomic bombings of Japan where thousands survived even though they suffered both radiation and burn injuries. But everyone wishes to provide the best care possible, even during an extreme emergency.

For over 70 years the American Red Cross has offered courses on simple nursing procedures that can be used in the home. Most communities have a program for training Emergency Medical Technicians (EMT's). This training would be very valuable. Existing EMT organizations, local health departments, and hospitals are usually able to provide courses on home nursing care, if enough people are interested. Professional nurses and physicians may be contacted to provide group instruction. There are undoubtedly other sources in most communities that could provide instruction.

It is recommended that as an absolute minimum the following nursing procedures should be thoroughly mastered:

—Handwashing technique.
—Taking body temperature.
—Taking pulse and respiration.
—Changing bed linen.
—Properly moving patients.
—Giving bed baths.
—Sanitary preparation and serving of food and fluids.
—Administration of medications.
—Oral hygiene.
—Skin care.
—Use of bedpans and urinals.
—Prevention of bedsores.
—Washing and sanitizing dishes, utensils, etc.
—Proper hot and cold pack application.
—Care of wounds, burns, lacerations.
—Inspecting the throat, ears, eyes, nose.
—Administration of ear drops, eye drops, nose drops.
—Management of bleeding.
—Burn care.
—Prevention of communicable diseases and infection.

REST

Often, the first order a physician will give a patient who shows signs of illness is to go to bed and rest. While at rest, patients are more comfortable, can keep quiet, warm, and generally more relaxed. During the course of an illness, people often sleep much more than usual. This is beneficial, since sleep allows the body to repair itself more rapidly.

In radiation sickness, complete rest must be ensured. It has been shown that exhaustive exercises following exposure to radiation enhance lethality. In nearly all instances where individuals have been exposed to full body doses of radiation of 200 rems and above, weakness and fatigue were present. Some patients recovering from doses of less than 350 rem have required 16 hours of rest daily, and tire after mild exertion. In severe radiation injury, after the toxic phase, patients may take months before their strength and endurance return to normal. Prostration is one of the initial symptoms of radiation injury in patients receiving intensive X-ray therapy. Weakness, fatigue, and prostration were characteristic of all patients involved in accidental radiation exposures at the Los Alamos Laboratories in the 1940s who received over 200 rem exposure. It was characteristic of the Japanese surviving radiation sickness after the Hiroshima and Nagasaki bombings.

Fatigue and weakness may show marked variability from person to person, and are often accompanied by nausea and vomiting in the initial phase of radiation sickness; but in some it is delayed for about 2 weeks after exposure. Complete rest is needed to better endure the toxic phase of radiation sickness. These periods of delayed weakness and fatigability correspond to the toxic (febrile) phase of the illness.

The foregoing represents a thorough program for the prevention and treatment of radiation sickness *on the family level*, including treatments and substances generally available to the public. Through a conscientious application of this program it is hoped that families can achieve a relative degree of preparedness that, coupled with other well-conceived measures, will give them a chance to go on. But it is important to note that this is only a part of preparedness planning and an informed awareness of vital nuclear issues. It is highly recommended that the following bibliography be examined, and the appropriate titles obtained.

APPENDIX

Appendix I

The table below lists the most commonly encountered radioisotopes, the radioactivity they give off (alpha, beta, gamma waves), and the half-lives of that activity. Since many of them give off more than one type of radiation, there is more than one period of half-life listed. Below, we discuss a few radioisotopes that are of particular interest because of their tendency to be taken up in the food chain, or for other reasons.

1. Radio-iodine 131 and 133: Because of the chemical identity of the radio-iodines with ordinary iodine they accumulate in the thyroid gland in humans and animals. Radio-iodine 131 has a half-life of eight days, and in the first sixty days following its introduction it will be the most hazardous source of internal radiation. Both of the radioactive iodines can be blocked from the thyroid gland through simple preventive measures. (See Chapter 9).

2. Cesium 137: This radioisotope, chemically very similar to potassium, usually is found collected in muscle tissues when consumed. Fortunately, it is not long retained in the body, being excreted as is potassium. It can cause severe cell damage, and has a half-life of 29.7 years.

3. Strontium 89 and 90: Often considered to be the greatest hazards of the radioisotopes. Strontium 89 and 90 chemically resemble calcium, an essential nutrient for humans, animals, and plants. Sr 89 has a half-life of only 50.5 days, while Sr 90 decays fifty percent in 27.7 years. The danger lies in the taking up of strontium radioisotopes into the food chain through plants and their subsequent use as food for humans and animals. Unlike cesium 137, it is not quickly passed through the body, and in even very small doses may cause problems, particularly bone cancer. As a result, young children and nursing mothers, whose bodies demand more calcium, are more likely to be affected. There is some evidence that calcium supplements taken prior to exposure *could* block the uptake of strontium, though it is not yet well-established.

4. Barium 140: This short-lived radioisotope (12.8 days) is quite similar to strontium 89 and 90, but because it decays so much more quickly, it constitutes less of a threat.

HALF-LIVES AND TYPE OF RADIATION ACTIVITY OF SOME RADIOISOTOPES

Isotope	Activity	Half-Life
Sodium 24	Beta, Gamma	14.9 hrs.
Potassium 40	Beta, Gamma	1.4×10^9 yrs.
Selenium 81	e^-, Beta	59 mins. — 17 mins.
Bromine 87	Beta, Gamma, Neutron	55.6 secs.
Krypton 92	Beta	2.4 secs.
Strontium 89	Beta	54 days
Strontium 90	Beta	25 years
Yttrium 93	Beta, Gamma	10 hours
Molybdenum 93	Beta, e^-, Gamma	6.7 hrs. — 2 yrs.
Technetium 99	Beta, Gamma, e^-	5.9 hrs — 5×10^5 yrs.
Ruthenium 106	Beta	1 year
Iodine 131	Beta, Gamma	8 days
Cesium 137	Beta, Gamma	33 years
Cerium 144	Alpha, Beta	290 days
Polonium 212	Alpha	3×10^7 secs.
Radium 226	Alpha, Beta	1,620 years
Actinium 227	Alpha, Beta, Gamma	27.7 years
Thorium 230	Alpha, Gamma	8×10^4 years
Thorium 232	Alpha, Gamma	1.39×10^{10} years
Protactinium 231	Alpha, Gamma	3.43×10^4 years
Uranium 234	Alpha, Gamma	2.48×10^5 years
Uranium 235	Alpha	7.1×10^8 years
Uranium 238	Alpha, Gamma	4.498×10^9 years
Uranium 239	Beta, Gamma	23.5 mins.
Neptunium 237	Alpha, Gamma	2.2×10^6 years
Neptunium 239	Beta, Gamma	2.3 days
Plutonium 238	Alpha, Gamma	92 years
Plutonium 239	Alpha, Gamma	2.41×10^4 years
Plutonium 240	Alpha	6.58×10^3 years
Plutonium 241	Alpha, Beta	14 years
Plutonium 242	Alpha	5×10^5 years
Americium 241	Alpha, Gamma	475 years
Americium 243	Alpha	10^4 years
Curium 243	Alpha	35 years
Curium 244	Alpha	19 years
Curium 245	Alpha	2×10^4 years
Curium 246	Alpha	3×10^3 years
Californium 249	Alpha	4.7×10^2 years
Californium 250	Alpha	12 years

Appendix II

DIRECTIONS FOR USING FORMS TO RECORD POSSIBLE SYMPTOMS OF RADIATION SICKNESS

It is necessary to carefully follow directions in recording any symptoms or conditions as a basis to determine if radiation sickness is present, and if so, its severity, treatment, and probable outcome.

Once radiation sickness occurs, it is necessary to treat the symptoms as outlined in the text. If the patient can survive the initial and toxic phases of the illness, the success of survival is very good. Recording should be done preferably in pencil.

1. Accurately fill out the appropriate spaces for name, date of birth, sex, address, and family physician.

2. When *any three* of the conditions or symptoms *occur at the same time* in an individual, begin recording for that patient as day one (1), then continue for the second day (day 2), and so on, *even if the symptoms or conditions no longer exist.*

In the columns marked S or M, across from each symptom or condition place an M if the problem is *mild*, place an S if the problem is *severe*. Whether the symptoms or conditions are mild or severe is a decision to be made by the person observing the patient who fills in the information.

Continue recording as long as any symptoms persist. If radiation sickness is present, the symptoms may subside, then reappear within 2 to 3 weeks, at which time the recording of symptoms should again be done.

One or even several of the symptoms occurring at the same time does not positively mean that radiation sickness exists. It may well be that anxiety, diet change, or other problems could be causing these symptoms, or they may indicate another illness totally unrelated to radiation sickness; but it will still be valuable to make a record.

SYMPTOMS AND CONDITIONS OF POSSIBLE RADIATION SICKNESS

Name: _____ Date of Birth: _____ Sex: ☐ Male
Address: _____ Family Physician: _____   ☐ Female

Symptom or Condition	Day:	1	2	3	4	5	6	7	8	9	10	11	12	13	14	15	16	17	18	19	20	21	22	23	24	25
	Date:																									
	Note:	Indicate severe (S) or mild (M) in boxes below for corresponding date.																								
Nausea																										
Diarrhea																										
Vomiting																										
Fever																										
Weakness																										
Loss of Appetite																										
Bleeding																										
Sore Mouth																										
Loss of Hair																										
Headache																										
Malaise																										
Infection																										
Skin Problems																										
Depression																										
Weight Loss																										
Extreme Thirst																										
Other																										

SYMPTOMS AND CONDITIONS OF POSSIBLE RADIATION SICKNESS

Name: _____ Date of Birth: _____ Sex: ☐ Male
Address: _____ Family Physician: _____   ☐ Female

Symptom or Condition	Day:	1	2	3	4	5	6	7	8	9	10	11	12	13	14	15	16	17	18	19	20	21	22	23	24	25
	Date:																									
	Note:	Indicate severe (S) or mild (M) in boxes below for corresponding date.																								
Nausea																										
Diarrhea																										
Vomiting																										
Fever																										
Weakness																										
Loss of Appetite																										
Bleeding																										
Sore Mouth																										
Loss of Hair																										
Headache																										
Malaise																										
Infection																										
Skin Problems																										
Depression																										
Weight Loss																										
Extreme Thirst																										
Other																										

NOTES

Part I
Chapter 1
[1] Herman Kahn, *On Thermonuclear War* (Princeton, NJ: Princeton University Press, 1960), p. 3.
[2] *Ibid.*, p. 15.
[3] Robert Scheer, "Pentagon Finishes Nuclear War Plan," *Salt Lake Tribune*, August 15, 1982, p. 1.

Chapter 2
[1] "Personal and Family Survival," *Civil Defense Adult Education Course Student Manual*, Office of Civil Defense, Department of Defense (Washington, DC: November 1966), p. iv.
[2] Howard Kornfeld, "Nuclear Weapons and Civil Defense," *The Western Journal of Medicine*, 138:2 (February, 1983), p. 211.
[3] Kahn, *op. cit.*, p. 18.

Chapter 3
[1] John Hersey, *Hiroshima* (New York: Bantam Books, 1946), pp. 60-61.
[2] Eisei Ishikawa and David L. Swain, translators, *Hiroshima and Nagasaki: The Physical, Medical, and Social Effects of the Atomic Bombings* (New York: Basic Books, 1981), p. 347.
[3] "Physician and Public Education on the Medical Consequences of Thermonuclear War," American Medical Assn. House of Delegates (Chicago: AMA Archives Report DD, 1981).
[4] "The Medical Consequences of Radiation Accidents and Nuclear War," American College of Physicians (Philadelphia, PA: Position Paper, April 16, 1982).
[5] R. J. Lifton and K. Erickson, "Nuclear War's Effect on the Mind," *The New York Times*, March 15, 1982, p. 14.
[6] Christine Cassel, "An Epistemology of Nuclear Weapons Effects," *The Western Journal of Medicine*, 138:2 (February, 1983), p. 213.
[7] *Ibid.*, p. 217.

Part II
Chapter 4
[1] Michael Riordan, editor, *The Day After Midnight* (Palo Alto, CA: Cheshire Books, 1982), p. 44.
[2] Kahn, *op. cit.*, pp. 130, 431.
[3] Jack Anderson, "EMP Could End World War III Before It Even Started," *The Deseret News*, December 3, 1982, Section A, p. 4.

Cresson Kearny, *Nuclear War Survival Skills* (Coos Bay, OR: Nuclear War Survival Research Bureau/Caroline House Publishers, Inc., 1982), p. 19.
[4]Ishikawa and Swain, *op. cit.*, p. 87.
[5]Los Alamos Scientific Laboratory, *The Effects of Nuclear Weapons* (Washington, DC: U.S. Government Printing Office: revised 1950, 1962, 1977; Samuel Glasstone and Philip J. Dolan, editors), p. 218.

Chapter 5
[1]Oliver Byrd, *Health* (Philadelphia: W.B. Saunders Co., 1961), p. 398.
[2]"Exposure to Radiation in an Emergency," National Committee on Radiation Protection and Measurements, NCRP Report No. 29, Washington, DC, 1962, pp. 38-39.

Chapter 6
[1]Riordan, *op. cit.*, pp. 101-102.
[2]"Protecting Our Food," *Yearbook of Agriculture* (Washington, DC: USDA, 1966), p. 340.
[3]*Ibid.*, pp. 341-344.
[4]Los Alamos Scientific Laboratory, *The Effects of Nuclear Weapons*: pp. 330-331.

Chapter 7
[1]"Exposure to Radiation in an Emergency," *op. cit.*, p. 35.
[2]*Effects of Nuclear Weapons*, pp. 322-23.
[3]Ben Freedman, *The Sanitarians Handbook* (New Orleans: Peerless Publishing Co., 1977), p. 1304.
[4]*The Effects of Nuclear Weapons*, pp. 321-322.
[5]*Ibid.*, p. 329.

Chapter 8
[1]William E. Deichmann and Horace Gerarde, *Symptomatology and Therapy of Toxicological Emergencies* (New York: Academic Press, 1964), p. 347.
[2]Lewis E. Etter, *The Science of Ionizing Radiation* (Springfield, IL: Charles C. Thomas, publisher, 1965), p. 232.
[3]Alexander Hollaender, *Radiation Biology* Vol. I, Part II (New York: McGraw-Hill, 1954), p. 909.
[4]Etter, *op. cit.*, p. 232.
[5]Hollaender, *op. cit.*, p. 920.
[6]*Ibid.*, p. 765.
[7]Louis H. Hempelmann, Herman Lisco and Joseph G. Hoffman, "The Acute Radiation Syndrome," *Annals of Internal Medicine* Volume 36, No. 2, Part I, February 1952, p. 385.
[8]*Ibid.*, p. 385.
[9]Etter, *op. cit.*, p. 237.
[10]Hempelmann, Lisco and Hoffman, *op. cit.*, p. 379.
[11]Richard E. King, "Survival in Nuclear Warfare," *CIBA Clinical Symposia* Vol. 14, No. 1, Jan.-Feb.-Mar., 1962, p. 29.
[12]Office of Civil Defense, *Family Guide Emergency Health Care*, 1961, p. 3.
[13]Hempelmann, Lisco and Hoffman, *op. cit.*, p. 287.

Chapter 9
[1]Hempelmann, Lisco and Hoffman, *op. cit.*, pp. 407-408.

²Michael G. Wohl and Robert S. Goodhard, *Modern Nutrition in Health and Disease* 3rd Edition (Philadelphia: Lea & Febiger, 1964), p. 883.
³H. M. Patt, E. B. Tyree, R. L. Straube and D. E. Smith, "Cysteine Protection Against X Irradiation," *Technical Papers, Science*, Vol. 110, August 26, 1949, p. 213.
⁴Hollaender, *op. cit.*, p. 943.
⁵*Ibid.*, p. 943.
⁶*Ibid.*
⁷*Ibid.*, p. 942.
⁸*Ibid.*
⁹*Ibid.*
¹⁰Kearny, *op. cit.*, p. 97.
¹¹"Final Recommendations: Potassium Iodide as a Thyroid-Blocking Agent in a Radiation Emergency," (U. S. Dept. of Health and Human Services, April, 1982), pp. 1-3.
¹²*Ibid.*, p. 4.
¹³*Ibid.*, p. 31.
¹⁴Kearny, *op. cit.*, p. 98.
¹⁵*Ibid.*, p. 99.
¹⁶"Final Recommendations: Potassium Iodide..." *op. cit.*, p. 18.
¹⁷*Ibid.*, p. 20.
¹⁸*Ibid.*, p. 22.
¹⁹Wohl and Goodhart, *op. cit.*, p. 440.
²⁰*Ibid.*, p. 441.
²¹*Ibid.*, p. 439.
²²"Clinical Experience of CGC Tablet (Cysteine, Vitamin C) in the Patients Treated with Radiation," *Nuclear Science Abstracts* #5319, Vol. 29, February, 1974.
²³J. Manowska (Jagiellonian University, Krakow), "Studies on the Protective Role of Vitamin C Under Conditions of Exposure of the Organisms to Ionizing Radiation," *Nuclear Science Abstracts* #29778, Vol. 33, June, 1976.

Chapter 10

¹Hollaender: p. 926; Etter: p. 239; Hempelmann, Lisco, Hoffman: p. 409; *The Merck Manual* 12th Edition (Merck Sharpe & Dohme Research Laboratories, 1972), pp. 1508-1509.
²Etter, *op. cit.*, p. 235.
³*Ibid.*, p. 238.
⁴David T. Smith and Norman F. Conant, *Microbiology*, 3rd Edition (New York: Appleton-Century-Crofts, 1960), p. 218.
⁵*Ibid.*
⁶Ishikawa and Swain, *op. cit.*, p. 535.
⁷*Physicians Desk Reference* 29th Edition (Oradell, NJ: Medical Economics Company, 1975), p. 866.
⁸*The Merck Manual, op. cit.*, p. 1508.
⁹Kahn, *op. cit.*, p. 86.
¹⁰Hempelmann, Lisco and Hoffman, *op. cit.*, p. 440.
¹¹*Ibid.*, p. 441.
¹²*Heinz Nutritional Data*, 6th Edition (H. J. Heinz Co., 1972), p. 4.
¹³*The Merck Manual, op. cit.*, p. 1495.

¹⁴Claudette Varricchio, "The Patient on Radiation Therapy," *American Journal of Nursing*, February, 1981, pp. 335-336.
¹⁵"Red Cross Home Nursing," (The American Red Cross, 1951), p. 29.
¹⁶*Ibid.*, p. 135.
¹⁷Richard King, *op. cit.*, p. 32.
¹⁸*Ibid.*, p. 32.

SELECTED BIBLIOGRAPHY

Books

Byrd, Oliver. *Health*. Philadelphia: W.B. Saunders Co., 1961.

Deichmann, William E. and Gerarde, Horace. *Symptomatology and Therapy of Toxicological Emergencies*. New York: Academic Press, 1964.

Etter, Lewis. *The Science of Ionizing Radiation*. Springfield, IL: Charles C. Thomas, Publisher, 1965.

Freedman, Ben. *The Sanitarians Handbook*. New Orleans: Peerless Publishing Co., 1977.

Glasstone, Samuel and Dolan, Philip J. *The Effects of Nuclear Weapons*. Washington, DC: U.S. Government Printing Office, Los Alamos Scientific Laboratory, revised 1950, 1962, 1977.

Hersey, John. *Hiroshima*. New York: Bantam Books, 1946.

Hollaender, Alexander. *Radiation Biology*. Vol. 1, Part II. New York: McGraw-Hill, 1954.

Ishikawa, Eisei and Swain, David L., translators. *Hiroshima and Nagasaki: The Physical, Medical, and Social Effects of the Atomic Bombings*. New York: Basic Books, 1981.

Kahn, Herman. *On Thermonuclear War*. Princeton, NJ: Princeton University Press, 1960.

Kearny, Cresson. *Nuclear War Survival Skills*. Coos Bay, OR: Nuclear War Survival Research Bureau/Caroline House Publishers, Inc., 1982.

Kennan, George F. *The Nuclear Delusion*. New York: Random House, 1982.

The Merck Manual. 12th Edition. Merck, Sharpe & Dohme Research Laboratories, 1972.

Physicians Desk Reference. 29th Edition. Oradell, NJ: Medical Economics Co., 1975.

Riordan, Michael. *The Day After Midnight*. Palo Alto, CA: Cheshire Books, 1982.

Salisbury, Harrison E. *The 900 Days*. New York: Harper & Row, 1969.

Scheer, Robert. *With Enough Shovels*. New York: Random House, 1982.

Schell, Jonathan. *The Fate of the Earth*. New York: Knopf, 1982.

BIBLIOGRAPHY 101

Smith, David T. and Conant, Norman F. *Microbiology*. 3rd Edition. New York: Appleton-Century-Crofts, 1960.

Wohl, Michael G. and Goodhard, Robert S. *Modern Nutrition in Health and Disease*. 3rd Edition. Philadelphia: Lea & Febiger, 1964.

Yearbook of Agriculture. Washington, DC: U.S. Department of Health and Human Services, 1966.

Papers

Cassel, Christine. "An Epistemology of Nuclear Weapons Effects," *The Western Journal of Medicine*. 138:2, February, 1983.

"Clinical Experience of CGC Tablet (Cysteine, Vitamin C) in the Patients Treated with Radiation," *Nuclear Science Abstracts*. #5319, Vol. 29, February, 1974.

"Exposure to Radiation in an Emergency." Washington, DC: National Committee on Radiation Protection and Measurements, 1962.

Family Guide Emergency Health Care. Washington, DC: Office of Civil Defense, 1961.

"Final Recommendations: Potassium Iodide as a Thyroid-Blocking Agent in a Radiation Emergency." Washington, DC: U.S. Department of Health and Human Services, April, 1962.

Heinz Nutritional Data, 6th Edition. H.J. Heinz Co., 1972.

King, Richard E. "Survival in Nuclear Warfare." CIBA Clinical Symposia Vol. 14. No. 1, Jan.-Feb.-Mar., 1962.

Hempelmann, Louis H.; Lisco, Herman and Hoffman, Joseph G. "The Acute Radiation Syndrome." *Annals of Internal Medicine*. Vol. 36, No. 2, Part I. February, 1952.

Kornfeld, Howard. "Nuclear Weapons and Civil Defense." *The Western Journal of Medicine*. 138:2. February, 1982.

Manowska, J. (Jagiellonian University, Krakow). "Studies on the Protective Role of Vitamin C Under Conditions of Exposure of the Organisms to Ionizing Radiation." *Nuclear Science Abstracts* #29778, Vol. 33, June, 1976.

"The Medical Consequences of Radiation Accidents and Nuclear War," Position Paper. Philadelphia, PA: American College of Physicians, April 16, 1982.

Patt, H.M.; Tyree, E.B.; Straube, R.L. and Smith, D.E. "Cysteine Protection Against X Irradiation," *Technical Papers, Science*. Vol. 110. August 26, 1949.

"Physician and Public Education on the Medical Consequences of Thermonuclear War." AMA Archives Report DD. Chicago, IL: American Medical Association House of Delegates, 1981.

"Red Cross Home Nursing," The American Red Cross, 1951.

Varricchio, Claudette. "The Patient on Radiation Therapy." *American Journal of Nursing*. February, 1981.